Lecture Notes in Mathematics

2101

T0225987

For further volumes:
http://www.springer.com/series/304

Saint-Flour Probability Summer School

The Saint-Flour volumes are reflections of the courses given at the Saint-Flour Probability Summer School. Founded in 1971, this school is organised every year by the Laboratoire de Mathématiques (CNRS and Université Blaise Pascal, Clermont-Ferrand, France). It is intended for PhD students, teachers and researchers who are interested in probability theory, statistics, and in their applications.

The duration of each school is 13 days (it was 17 days up to 2005), and up to 70 participants can attend it. The aim is to provide, in three high-level courses, a comprehensive study of some fields in probability theory or Statistics. The lecturers are chosen by an international scientific board. The participants themselves also have the opportunity to give short lectures about their research work.

Participants are lodged and work in the same building, a former seminary built in the 18th century in the city of Saint-Flour, at an altitude of 900 m. The pleasant surroundings facilitate scientific discussion and exchange.

The Saint-Flour Probability Summer School is supported by:

– Université Blaise Pascal
– Centre National de la Recherche Scientifique (C.N.R.S.)
– Ministère délégué à l'Enseignement supérieur et à la Recherche

For more information, see back pages of the book and
http://math.univ-bpclermont.fr/stflour/

Jean Picard
Summer School Chairman
Laboratoire de Mathématiques
Université Blaise Pascal
63177 Aubière Cedex
France

Takashi Kumagai

Random Walks on Disordered Media and their Scaling Limits

École d'Été de Probabilités
de Saint-Flour XL - 2010

 Springer

Takashi Kumagai
Research Institute
    for Mathematical Sciences
Kyoto University
Kyoto, Japan

ISBN 978-3-319-03151-4   ISBN 978-3-319-03152-1 (eBook)
DOI 10.1007/978-3-319-03152-1
Springer Cham Heidelberg New York Dordrecht London

Lecture Notes in Mathematics ISSN print edition: 0075-8434
                                ISSN electronic edition: 1617-9692

Library of Congress Control Number: 2013957872

Mathematics Subject Classification (2010): 60-xx, 35Qxx, 31-xx

Printed on acid-free paper

Springer is part of Springer Science+Business Media (www.springer.com)

Preface

The main theme of these lecture notes is to analyze heat conduction on disordered media such as fractals and percolation clusters by means of both probabilistic and analytic methods, and to study the scaling limits of Markov chains on the media.

The problem of random walk on a percolation cluster "the ant in the labyrinth" has received much attention both in the physics and in the mathematics literature. In 1986, H. Kesten showed an anomalous behavior of a random walk on a percolation cluster at critical probability for trees and for $\mathbb{Z}^2$. (To be precise, the critical percolation cluster is finite, so the random walk is considered on an incipient infinite cluster (IIC), namely a critical percolation cluster conditioned to be infinite.) Partly motivated by this work, analysis and diffusion processes on fractals have been developed since the late 1980s. As a result, various new methods have been produced to estimate heat kernels on disordered media, and these turn out to be useful to establish quenched estimates on random media. Recently, it has been proved that random walks on IICs are sub-diffusive on $\mathbb{Z}^d$ when d is high enough, on trees, and on the spread-out percolation for $d > 6$.

Throughout the lecture notes, I will survey the above-mentioned developments in a compact way. In the first part, I will summarize some classical and nonclassical estimates for heat kernels and discuss stability of the estimates under perturbations of operators and spaces. Here Nash inequalities and equivalent inequalities will play a central role. In the latter part, I will give various examples of disordered media and obtain heat kernel estimates for Markov chains on them. In some models, I will also discuss scaling limits of the Markov chains. Examples of disordered media include fractals, percolation clusters, random conductance models, and random graphs.

In summer 2013, I have made a revised final version of the notes. There has been significant progress in the areas for the 3 years after my St. Flour lectures. I have tried to include the information of recent developments as much as possible, but they are clearly far from complete. I also put several new ingredients concerning heat kernel estimates (Sects. 3.3 and 3.4).

For typos and errors, I will update a list of corrections at the following page.

http://www.kurims.kyoto-u.ac.jp/~kumagai/

Kyoto, Japan   Takashi Kumagai
September 2013

Acknowledgements

These are notes from a series of eight lectures given at the Saint-Flour Probability Summer School, July 4–17, 2010. Part of them was also lectured in the first NIMS summer school in probability 2011 at Daejeon, Korea, Series of lectures 2011 at Kobe University, Japan, and Spring School in Probability 2012 at IUC, Dubrovnik, Croatia. I would like to express my gratitude to the organizers, J. Picard, K.-H. Kim, P. Kim, S. Lee, K. Fukuyama, Y. Higuchi, R. Schilling, R. Song, and Z. Vondraček, for their hospitality during the events.

I am very grateful to M.T. Barlow, M. Biskup, Z.-Q. Chen, J.-D. Deuschel, J. Kigami, G. Kozma, P. Mathieu, A. Nachmias, H. Osada, A. Sapozhnikov, T. Shirai, G. Slade, H. Tanemura, R. van der Hofstad, and O. Zeitouni for fruitful discussions and comments, and G. Ben Arous, E. Bolthausen, F. den Hollander, Y. Peres, and A.-S. Sznitman for their encouragements. Especially, very detailed comments from M. Biskup, G. Kozma, and A. Sapozhnikov to the first draft were extremely helpful to improve the notes. I would also like to thank the audience of lectures including the audience of pre-lectures in Kyoto, especially Y. Abe, E. Baur, D. Croydon, M. Felsinger, R. Fitzner, R. Fukushima, N. Kajino, N. Kubota, K. Kuwada, M. Kwaśnicki, J.C. Mourrat, M. Nakashima, H. Nawa, G. Pete, M. Sasada, M. Shinoda, D. Shiraishi, P. Sosoe, M. Takei, and N. Yoshida for useful comments and for finding typos in the draft. (To be precise, some of them were not the audience of lectures, but sent me comments/corrections by emails.)

The final corrections of the notes are made when I was visiting Max Planck Institute in Leipzig and Universität Leipzig. I would like to thank Max von Renesse and A. Sapozhnikov for their hospitality during my stay.

This work was supported in part by the Grant-in-Aid for Scientific Research (B) 22340017 and (A) 25247007.

Contents

Chapter 1
Introduction

Around mid-1960s, mathematical physicists began to analyze properties of disordered media such as structures of polymers and networks, growth of molds and crystals (see for example, [44]). They observed anomalous behavior for random walks/diffusions on the media. One of the typical examples of disordered media is a percolation cluster. To be precise, the bond percolation model on the lattice $\mathbb{Z}^d$ is defined as follows: each nearest neighbor bond is open with probability $p \in (0, 1)$ and closed otherwise, independently of all the others. Let $\mathcal{C}(x)$ be the open cluster containing x. Then if $\theta(p) := \mathbb{P}_p(|\mathcal{C}(x)| = \infty)$ it is well known (see [122]) that for each $d \geq 2$, there exists $p_c = p_c(d) \in (0, 1)$ such that $\theta(p) = 0$ if $p < p_c$ and $\theta(p) > 0$ if $p > p_c$. For $p > p_c$, there exists unique ∞-cluster (i.e. an open connected component whose size is infinity). In the pioneering work, de Gennes [110] pondered the problem of a random walker in percolation clusters, which he described as "ant in the labyrinth". In 1982, Alexander and Orbach [8] made a stimulating conjecture concerning the anomalous behavior of the random walk on the percolation cluster at p_c. Let $Y = \{Y_n^\omega\}_{n\in\mathbb{N}}$ be the simple random walk on the cluster, and $p_n^\omega(x, y)$ be its heat kernel. Define

$$d_s := -2 \lim_{n\to\infty} \frac{\log p_{2n}^\omega(x, x)}{\log n} \tag{1.1}$$

as the spectral dimension of the cluster (if the limit exists). The Alexander–Orbach conjecture says $d_s = 4/3$ for all $d \geq 2$. (Note that $d_s = d$ for simple random walk on $\mathbb{Z}^d$.)

Mathematical challenges towards these problems started late 1980s. In 1986, Kesten wrote two beautiful papers [149,150] for random walk on critical percolation clusters conditioned not to extinct for trees and for $\mathbb{Z}^2$. These are the first mathematical rigorous work in this direction, and they triggered intensive research on diffusions on fractals and analysis on fractals [16, 151]. After a few decades, the area is mature enough to obtain rigorous results for the original problems on disordered media.

T. Kumagai, *Random Walks on Disordered Media and their Scaling Limits*, Lecture Notes in Mathematics 2101, DOI 10.1007/978-3-319-03152-1_1,
© Springer International Publishing Switzerland 2014

In these lecture notes, we will survey the above mentioned developments in a compact way. We will focus on the discrete potential theory and how the theory is effectively used in the analysis of disordered media.

Chapter 2 is an introduction to general potential theory for symmetric Markov chains on weighted graphs. We then show various equivalent/sufficient conditions for the heat kernel upper bounds (the Nash inequality), and for sub-Gaussian heat kernel bounds in Chap. 3. In Chap. 4, we use effective resistance to estimate Green functions, exit times from balls etc. On-diagonal heat kernel bounds are also obtained using the effective resistance.

Chapters 5–7 are for the random walk on an incipient infinite cluster (IIC) for a critical percolation. In Chap. 5, we give some sufficient condition for the sharp on-diagonal heat kernel bounds for random walk on random graphs. We then prove in Chap. 6 the Alexander–Orbach conjecture for IICs when two-point functions behave nicely, especially for IICs of high dimensional critical bond percolations on $\mathbb{Z}^d$. We also discuss heat kernel bounds and scaling limits on related random models in Chap. 7. Chapter 8 will be devoted to the quenched invariance principle for the random conductance model on $\mathbb{Z}^d$. Put i.i.d. conductance μ_e on each bond in $\mathbb{Z}^d$ and consider the Markov chain associated with the (random) weighted graph. We consider two cases, namely $0 \le \mu_e \le c$ $\mathbb{P}$-a.e. and $c \le \mu_e < \infty$ $\mathbb{P}$-a.e. Although the behavior of heat kernels are quite different, for both cases (and for a general case $0 \le \mu_e < \infty$ $\mathbb{P}$-a.e.) the scaling limit is Brownian motion in general. We discuss some technical details about correctors of the Markov chains, which play a key role to prove the invariance principle.

We referred many papers and books during the preparations of the lecture notes. Especially, we owe a lot to the lecture notes by Barlow [20] and by Coulhon [75] for Chaps. 2 and 3. Chapters 2–5 (and part of Chap. 7) are mainly from [32, 166]. Chapter 6 is mainly from a beautiful paper by Kozma and Nachmias [162] (with some simplification in [197]). In Chap. 8, we follow the arguments of the papers by Barlow, Biskup, Deuschel, Mathieu and their co-authors [28, 51, 52, 59, 178, 179]. The lecture notes by Biskup [55] is the best reference to overview the area and its history.

Throughout the lecture notes, we use c, c' to denote strictly positive finite constants whose values are not significant and may change from line to line. In principle, we write c_i, C_i for positive constants whose values are fixed within theorems and lemmas. Occasionally, we adopt the convention that if we cite elsewhere the constant c_1 in Lemma 2.1.2 (for example), we denote it as $c_{2.1.2.1}$.

For two real numbers a and b, $a \wedge b := \min\{a, b\}$ and $a \vee b := \max\{a, b\}$. For functions f and g, $f \asymp g$ means that there exist constants $0 < c_1 \le c_2$ such that $c_1 g(x) \le f(x) \le c_2 g(x)$ for all x. For functions f and g in $\mathbb{R}^d$, $f \sim g$ means $f(x)/g(x) \to 1$ as $|x| \to \infty$.

Chapter 2
Weighted Graphs and the Associated Markov Chains

In this chapter, we discuss general potential theory for symmetric (reversible) Markov chains on weighted graphs. Note that there are many nice books and lecture notes that treat potential theory and/or Markov chains on graphs, for example [7, 20, 93, 118, 125, 175, 195, 204, 211]. While writing this chapter, we are largely influenced by the lecture notes by Barlow [20].

2.1 Weighted Graphs

Let X be a finite or a countably infinite set, and E is a subset of $\{\{x, y\} : x, y \in X, x \neq y\}$. A graph is a pair (X, E). For $x, y \in X$, we write $x \sim y$ if $\{x, y\} \in E$. A sequence $x_0, x_1, \cdots, x_n$ is called a path with length n if $x_i \in X$ for $i = 0, 1, 2, \cdots, n$ and $x_j \sim x_{j+1}$ for $j = 0, 1, 2, \cdots, n - 1$. For $x \neq y$, define $d(x, y)$ to be the length of the shortest path from x to y. If there is no such path, we set $d(x, y) = \infty$ and we set $d(x, x) = 0$. $d(\cdot, \cdot)$ is a metric on X and it is called a graph distance. (X, E) is connected if $d(x, y) < \infty$ for all $x, y \in X$, and it is locally finite if $|\{y : \{x, y\} \in E\}| < \infty$ for all $x \in X$. Throughout the lectures, we will consider connected locally finite graphs (except when we consider the trace of them in Sect. 2.3).

Assume that the graph (X, E) is endowed with a weight (conductance) μ_{xy}, which is a symmetric nonnegative function on $X \times X$ such that $\mu_{xy} > 0$ if and only if $x \sim y$. We call the pair (X, μ) a weighted graph.

Let $\mu_x = \mu(x) = \sum_{y \in X} \mu_{xy}$ and define a measure μ on X by setting $\mu(A) = \sum_{x \in A} \mu_x$ for $A \subset X$. Also, we define $B(x, r) = \{y \in X : d(x, y) < r\}$ for each $x \in X$ and $r \geq 1$.

Definition 2.1.1. We say that (X, μ) has controlled weights (or (X, μ) satisfies p_0-condition) if there exists $p_0 > 0$ such that

T. Kumagai, *Random Walks on Disordered Media and their Scaling Limits*, Lecture Notes in Mathematics 2101, DOI 10.1007/978-3-319-03152-1_2,
© Springer International Publishing Switzerland 2014

$$\frac{\mu_{xy}}{\mu_x} \geq p_0 \qquad \forall x \sim y.$$

If (X, μ) has controlled weights, then clearly $|\{y \in X : x \sim y\}| \leq p_0^{-1}$.

Once the weighted graph (X, μ) is given, we can define the corresponding quadratic form, Markov chain and the discrete Laplace operator.

Quadratic Form. We define a quadratic form on (X, μ) as follows.

$$H^2(X, \mu) = H^2 = \{f : X \to \mathbb{R} : \mathcal{E}(f, f) = \frac{1}{2} \sum_{\substack{x, y \in X \\ x \sim y}} (f(x) - f(y))^2 \mu_{xy} < \infty\},$$

$$\mathcal{E}(f, g) = \frac{1}{2} \sum_{\substack{x, y \in X \\ x \sim y}} (f(x) - f(y))(g(x) - g(y)) \mu_{xy} \qquad \forall f, g \in H^2.$$

Physically, $\mathcal{E}(f, f)$ is the energy (per unit time) of the electrical network for an (electric) potential f.

Since the graph is connected, one can easily see that $\mathcal{E}(f, f) = 0$ if and only if f is a constant function. We fix a base point $0 \in X$ and define

$$\|f\|_{H^2}^2 = \mathcal{E}(f, f) + f(0)^2 \qquad \forall f \in H^2.$$

Note that

$$\mathcal{E}(f, f) = \frac{1}{2} \sum_{x \sim y} (f(x) - f(y))^2 \mu_{xy} \leq \sum_x \sum_y (f(x)^2 + f(y)^2) \mu_{xy} = 2\|f\|_2^2,$$

(2.1)

for all $f \in \mathbb{L}^2$ where $\|f\|_2 := (\sum_x f(x)^2 \mu_x)^{1/2}$ is the $\mathbb{L}^2$-norm of f. So $\mathbb{L}^2 \subset H^2$. We give basic facts in the next lemma.

Lemma 2.1.2. *(i) Convergence in H^2 implies the pointwise convergence.*
(ii) H^2 is a Hilbert space.

Proof. (i) Suppose $f_n \to f$ in H^2 and let $g_n = f_n - f$. Then $\mathcal{E}(g_n, g_n) + g_n(0)^2 \to 0$ so $g_n(0) \to 0$. For any $x \in X \setminus \{0\}$, there is a sequence $\{x_i\}_{i=0}^l \subset X$ such that $x_0 = 0, x_l = x, x_i \sim x_{i+1}$ for $i = 0, 1, \cdots, l - 1$ and $x_i \neq x_j$ for $i \neq j$. Then

$$|g_n(x) - g_n(0)|^2 \leq l \sum_{i=0}^{l-1} |g_n(x_i) - g_n(x_{i+1})|^2 \leq 2l (\min_{i=0}^{l-1} \mu_{x_i x_{i+1}})^{-1} \mathcal{E}(g_n, g_n) \to 0$$

(2.2)

as $n \to \infty$ so we have $g_n(x) \to 0$.

(ii) Assume that $\{f_n\}_n \subset H^2$ is a Cauchy sequence in H^2. Then $f_n(0)$ is a Cauchy sequence in $\mathbb{R}$ so converges. Thus, similarly to (2.2) f_n converges pointwise to f, say. Now using Fatou's lemma, we have $\| f_n - f \|_{H^2}^2 \leq \liminf_m \| f_n - f_m \|_{H^2}^2$, so that $\| f_n - f \|_{H^2}^2 \to 0$. $\qquad\qquad\qquad\qquad\qquad\qquad\qquad\qquad\qquad\quad\square$

Markov Chain. Let $Y = \{Y_n\}$ be a Markov chain on X whose transition probabilities are given by

$$\mathbb{P}(Y_{n+1} = y | Y_n = x) = \frac{\mu_{xy}}{\mu_x} =: P(x, y) \qquad \forall x, y \in X.$$

We write $\mathbb{P}^x$ when the initial distribution of Y is concentrated on x (i.e. $Y_0 = x$, $\mathbb{P}$-a.s.). $(P(x, y))_{x,y \in X}$ is the transition matrix for Y. Y is called a simple random walk when $\mu_{xy} = 1$ whenever $x \sim y$. Y is μ-symmetric since for each $x, y \in X$,

$$\mu_x P(x, y) = \mu_{xy} = \mu_{yx} = \mu_y P(y, x).$$

We define the *heat kernel* of Y by

$$p_n(x, y) = \mathbb{P}^x(Y_n = y)/\mu_y \qquad \forall x, y \in X. \tag{2.3}$$

Using the Markov property, we can easily show the Chapman-Kolmogorov equation:

$$p_{n+m}(x, y) = \sum_z p_n(x, z) p_m(z, y) \mu_z, \qquad \forall x, y \in X. \tag{2.4}$$

Using this and the fact $p_1(x, y) = \mu_{xy}/(\mu_x \mu_y) = p_1(y, x)$, one can verify the following inductively

$$p_n(x, y) = p_n(y, x), \qquad \forall x, y \in X.$$

For $n \geq 1$ and $f : X \to \mathbb{R}$, let

$$P_n f(x) = \sum_y p_n(x, y) f(y) \mu_y = \sum_y \mathbb{P}^x(Y_n = y) f(y) = \mathbb{E}^x[f(Y_n)].$$

We sometimes consider a continuous time Markov chain $\{Y_t\}_{t \geq 0}$ with respect to μ which is defined as follows: each particle stays at a point, say x for (independent) exponential time with parameter 1, and then jumps to another point, say y with probability $P(x, y)$. The heat kernel for the continuous time Markov chain can be expressed as follows.

$$p_t(x, y) = \mathbb{P}^x(Y_t = y)/\mu_y = \sum_{n=0}^{\infty} e^{-t} \frac{t^n}{n!} p_n(x, y), \qquad \forall x, y \in X.$$

Discrete Laplace Operator. For $f : X \to \mathbb{R}$, the discrete Laplace operator is defined by

$$\mathcal{L}f(x) = \sum_y P(x, y)f(y) - f(x) = \frac{1}{\mu_x} \sum_y (f(y) - f(x))\mu_{xy}$$

$$= \mathbb{E}^x[f(Y_1)] - f(x) = (P_1 - I)f(x), \tag{2.5}$$

where Y_1 is the (discrete time) Markov chain on X at time 1. Note that according to Ohm's law "$I = V/R$", $\sum_y (f(y) - f(x))\mu_{xy}$ is the total flux flowing into x, given the potential f.

Definition 2.1.3. Let $A \subset X$. A function $f : X \to \mathbb{R}$ is harmonic on A if

$$\mathcal{L}f(x) = 0, \qquad \forall x \in A.$$

f is sub-harmonic (resp. super-harmonic) on A if $\mathcal{L}f(x) \geq 0$ (resp. $\mathcal{L}f(x) \leq 0$) for $x \in A$.

$\mathcal{L}f(x) = 0$ means that the total flux flowing into x is 0 for the given potential f. This is the behavior of the currents in a network called Kirchhoff's (first) law.

For $A \subset X$, we define the (exterior) boundary of A by

$$\partial A = \{x \in A^c : \exists z \in A \text{ such that } z \sim x\}. \tag{2.6}$$

Proposition 2.1.4 (Maximum Principle). *Let A be a connected subset of X and $h : A \cup \partial A \to \mathbb{R}$ be sub-harmonic on A. If the maximum of h over $A \cup \partial A$ is attained in A, then h is constant on $A \cup \partial A$.*

Proof. Let $x_0 \in A$ be the point where h attains the maximum and let $H = \{z \in A \cup \partial A : h(z) = h(x_0)\}$. If $y \in H \cap A$, then since $h(y) \geq h(x)$ for all $x \in A \cup \partial A$, we have

$$0 \leq \mu_y \mathcal{L}h(y) = \sum_x (h(x) - h(y))\mu_{xy} \leq 0.$$

Thus, $h(x) = h(y)$ (i.e. $x \in H$) for all $x \sim y$. Since A is connected, this implies $H = A \cup \partial A$. $\qquad \square$

We can prove the minimum principle for a super-harmonic function h by applying the maximum principle to $-h$.

For $f, g \in \mathbb{L}^2$, denote their $\mathbb{L}^2$-inner product as (f, g), namely $(f, g) = \sum_x f(x)g(x)\mu_x$.

Lemma 2.1.5. *(i) $\mathcal{L} : H^2 \to \mathbb{L}^2$ and $\|\mathcal{L}f\|_2^2 \leq 2\|f\|_{H^2}^2$.*
(ii) For $f \in H^2$ and $g \in \mathbb{L}^2$, we have $(-\mathcal{L}f, g) = \mathcal{E}(f, g)$.
(iii) $\mathcal{L}$ is a self-adjoint operator on $\mathbb{L}^2(X, \mu)$ and the following holds:

$$(-\mathcal{L}f, g) = (f, -\mathcal{L}g) = \mathcal{E}(f, g), \qquad \forall f, g \in \mathbb{L}^2. \tag{2.7}$$

Proof. (i) Using the Schwarz inequality, we have

$$\|\mathcal{L}f\|_2^2 = \sum_x \frac{1}{\mu_x}(\sum_y (f(y) - f(x))\mu_{xy})^2$$

$$\leq \sum_x \frac{1}{\mu_x}(\sum_y (f(y) - f(x))^2\mu_{xy})(\sum_y \mu_{xy}) = 2\mathcal{E}(f, f) \leq 2\|f\|_{H^2}^2.$$

(ii) Using (i), both sides of the equality are well-defined. Further, using the Schwarz inequality,

$$\sum_{x,y}|\mu_{xy}(f(y) - f(x))g(x)| \leq (\sum_{x,y}\mu_{xy}(f(y) - f(x))^2)^{1/2}(\sum_{x,y}\mu_{xy}g(x)^2)^{1/2}$$

$$= \sqrt{2}\mathcal{E}(f, f)^{1/2}\|g\|_2 < \infty.$$

So we can use Fubini's theorem, and we have

$$(-\mathcal{L}f, g) = -\sum_x(\sum_y \mu_{xy}(f(y) - f(x)))g(x)$$

$$= \frac{1}{2}\sum_x \sum_y \mu_{xy}(f(y) - f(x))(g(y) - g(x)) = \mathcal{E}(f, g).$$

(iii) We can prove $(f, -\mathcal{L}g) = \mathcal{E}(f, g)$ similarly and obtain (2.7). $\quad\square$

Equation (2.7) is the discrete Gauss-Green formula.

Lemma 2.1.6. *Set $p_n^x(\cdot) = p_n(x, \cdot)$. Then, the following hold for all $x, y \in X$.*

$$p_{n+m}(x, y) = (p_n^x, p_m^y), \quad P_1 p_n^x(y) = p_{n+1}^x(y), \tag{2.8}$$

$$\mathcal{L}p_n^x(y) = p_{n+1}^x(y) - p_n^x(y), \quad \mathcal{E}(p_n^x, p_m^y) = p_{n+m}^x(y) - p_{n+m+1}^x(y), \tag{2.9}$$

$$p_{2n}(x, y) \leq \sqrt{p_{2n}(x, x)p_{2n}(y, y)}. \tag{2.10}$$

Proof. The two equations in (2.8) are due to the Chapman-Kolmogorov equation (2.4). The first equation in (2.9) is then clear since $\mathcal{L} = P_1 - I$. The second equation in (2.9) can be obtained by these equations and (2.7). Using (2.8) and the Schwarz inequality, we have

$$p_{2n}(x, y)^2 = (p_n^x, p_n^y)^2 \leq (p_n^x, p_n^x)(p_n^y, p_n^y) = p_{2n}(x, x)p_{2n}(y, y),$$

which gives (2.10). $\quad\square$

It can be easily shown that $(\mathcal{E}, \mathbb{L}^2)$ is a regular Dirichlet form on $\mathbb{L}^2(X, \mu)$ (cf. [108]). Then the corresponding Hunt process is the continuous time Markov chain $\{Y_t\}_{t \geq 0}$ with respect to μ and the corresponding self-adjoint operator on $\mathbb{L}^2$ is $\mathcal{L}$ in (2.5).

Remark 2.1.7. Note that $\{Y_t\}_{t \geq 0}$ has the transition probability $P(x, y) = \mu_{xy}/\mu_x$ and it waits at x for an exponential time with mean 1 for each $x \in X$. Since the "speed" of $\{Y_t\}_{t \geq 0}$ is independent of the location, it is sometimes called constant speed random walk (CSRW for short). We can also consider a continuous time Markov chain with the same transition probability $P(x, y)$ and wait at x for an exponential time with mean μ_x^{-1} for each $x \in X$. This Markov chain is called variable speed random walk (VSRW for short). We will discuss VSRW in Chap. 8. The corresponding discrete Laplace operator is

$$\mathcal{L}_V f(x) = \sum_y (f(y) - f(x))\mu_{xy}. \tag{2.11}$$

For each f, g that have finite support, we have

$$\mathcal{E}(f, g) = -(\mathcal{L}_V f, g)_\nu = -(\mathcal{L} f, g)_\mu,$$

where ν is a measure on X such that $\nu(A) = |A|$ for all $A \subset X$. So VSRW is the Markov process associated with the Dirichlet form $(\mathcal{E}, \mathbb{L}^2)$ on $\mathbb{L}^2(X, \nu)$ and CSRW is the Markov process associated with the Dirichlet form $(\mathcal{E}, \mathbb{L}^2)$ on $\mathbb{L}^2(X, \mu)$. VSRW is a time changed process of CSRW and vice versa.

We now introduce the notion of rough isometry.

Definition 2.1.8. Let $(X_1, \mu_1), (X_2, \mu_2)$ be weighted graphs that have controlled weights.

(i) A map $T : X_1 \rightarrow X_2$ is called a rough isometry if the following holds. There exist constants $c_1, c_2, c_3 > 0$ such that

$$c_1^{-1} d_1(x, y) - c_2 \leq d_2(T(x), T(y)) \leq c_1 d_1(x, y) + c_2 \ \forall x, y \in X_1, \tag{2.12}$$

$$d_2(T(X_1), y') \leq c_2 \quad \forall y' \in X_2, \tag{2.13}$$

$$c_3^{-1} \mu_1(x) \leq \mu_2(T(x)) \leq c_3 \mu_1(x) \quad \forall x \in X_1, \tag{2.14}$$

where $d_i(\cdot, \cdot)$ is the graph distance of (X_i, μ_i), for $i = 1, 2$.

(ii) $(X_1, \mu_1), (X_2, \mu_2)$ are said to be rough isometric if there is a rough isometry between them.

It is easy to see that rough isometry is an equivalence relation. One can easily prove that $\mathbb{Z}^2$, the triangular lattice, and the hexagonal lattice are all roughly isometric if there exists $M > 0$ such that $\mu_{xy} \in [M^{-1}, M]$ whenever $x \sim y$. It can be proved that $\mathbb{Z}^1$ and $\mathbb{Z}^2$ are not roughly isometric.

The notion of rough isometry was first introduced by M. Kanai [145, 146]. Since his work was mainly concerned with Riemannian manifolds, definition of rough isometry included only (2.12), (2.13). The definition equivalent to Definition 2.1.8 is given in [77] (see also [130]). Note that rough isometry corresponds to (coarse) quasi-isometry in the field of geometric group theory, which was introduced by Gromov already in 1981 (see [124]).

When we discuss various properties of Markov chains/Laplace operators, it is important to think about their "stability". In the following, we introduce two types of stability.

Definition 2.1.9.   (i) We say a property is stable under bounded perturbation if whenever (X, μ) satisfies the property and (X, μ') satisfies $c^{-1}\mu_{xy} \le \mu'_{xy} \le c\mu_{xy}$ for all $x, y \in X$, then (X, μ') satisfies the property.
(ii)   We say a property is stable under rough isometry if whenever (X, μ) satisfies the property and (X', μ') is rough isometric to (X, μ), then (X', μ') satisfies the property.

If a property is stable under rough isometry, then it is clearly stable under bounded perturbation.

It is known that the following properties of weighted graphs are stable under rough isometry.

(i)   Transience and recurrence
(ii)   The Nash inequality, i.e. $p_n(x, y) \le c_1 n^{-\alpha}$ for all $n \ge 1, x, y \in X$ (for some $\alpha > 0$)
(iii)   Parabolic Harnack inequality (see Definition 3.3.4 (2))

We will see (i) later in this chapter, and (ii) and (iii) in Chap. 3. One of the important open problems is to show if the elliptic Harnack inequality, i.e. the Harnack inequality for harmonic functions, is stable under these perturbations or not. (In fact, recently this has been affirmatively solved in [38] under some assumption. Yet, the assumption contains some regularity of the growth of capacities and occupation times. It would be desirable to prove (or disprove) the stability assuming only the volume growth condition.)

Definition 2.1.10. (X, μ) has the Liouville property if there is no bounded non-constant harmonic functions. (X, μ) has the strong Liouville property if there is no positive non-constant harmonic functions.

It is known that both Liouville and strong Liouville properties are not stable under bounded perturbation (see [176], also [45] for a counterexample in the framework of graphs/manifolds with polynomial volume growth).

2.2   Harmonic Functions and Effective Resistances

For $A \subset X$, define

$$\sigma_A = \inf\{n \ge 0 : Y_n \in A\}, \quad \sigma_A^+ = \inf\{n > 0 : Y_n \in A\},$$

$$\tau_A = \inf\{n \ge 0 : Y_n \notin A\}.$$

For $A \subset X$ and $f : A \to \mathbb{R}$, consider the following *Dirichlet problem*.

$$\begin{cases} \mathcal{L}v(x) = 0 \quad \forall x \in A^c, \\ v|_A = f. \end{cases} \tag{2.15}$$

Proposition 2.2.1. *Assume that $f : A \to \mathbb{R}$ is bounded and set*

$$\varphi(x) = \mathbb{E}^x[f(Y_{\sigma_A}) : \sigma_A < \infty].$$

(i) φ *is a solution of* (2.15).
(ii) *If* $\mathbb{P}^x(\sigma_A < \infty) = 1$ *for all* $x \in X$, *then* φ *is the unique bounded solution of* (2.15).

Proof. (i) $\varphi|_A = f$ is clear. For $x \in A^c$, using the Markov property of Y, we have

$$\varphi(x) = \sum_y P(x, y)\varphi(y),$$

so $\mathcal{L}\varphi(x) = 0$.
(ii) Let φ' be another bounded solution and let $H_n = \varphi(Y_n) - \varphi'(Y_n)$. Then H_n is a bounded martingale up to σ_A, so using the optional stopping theorem, we have

$$\varphi(x) - \varphi'(x) = \mathbb{E}^x H_0 = \mathbb{E}^x H_{\sigma_A} = \mathbb{E}^x[\varphi(Y_{\sigma_A}) - \varphi'(Y_{\sigma_A})]$$
$$= \mathbb{E}^x[f(Y_{\sigma_A}) - f(Y_{\sigma_A})] = 0$$

since $\sigma_A < \infty$ a.s. and $\varphi(x) = \varphi'(x)$ for $x \in A$.   □

Remark 2.2.2. (i) In particular, we see that φ is the unique solution of (2.15) when A^c is finite. In this case, we have another proof of the uniqueness of the solution of (2.15): let $u(x) = \varphi(x) - \varphi'(x)$, then $u|_A = 0$ and $\mathcal{L}u(x) = 0$ for $x \in A^c$. So, noting $u \in \mathbb{L}^2$ and using Lemma 2.1.5, $\mathcal{E}(u, u) = (-\mathcal{L}u, u) = 0$ which implies that u is constant on X (so it is 0 since $u|_A = 0$).
(ii) If $h_A(x) := \mathbb{P}^x(\sigma_A = \infty) > 0$ for some $x \in X$, then the function $\varphi + \lambda h_A$ is also a solution of (2.15) for all $\lambda \in \mathbb{R}$, so the uniqueness of the Dirichlet problem fails.

For $A, B \subset X$ such that $A \cap B = \emptyset$, define

$$R_{\mathrm{eff}}(A, B)^{-1} = \inf\{\mathcal{E}(f, f) : f \in H^2, f|_A = 1, f|_B = 0\}. \tag{2.16}$$

(We define $R_{\mathrm{eff}}(A, B) = \infty$ when the right hand side is 0, and $R_{\mathrm{eff}}(A, B) = 0$ when there is no $f \in H^2$ that satisfies $f|_A = 1$ and $f|_B = 0$.) We call $R_{\mathrm{eff}}(A, B)$ the *effective resistance* between A and B. It is easy to see that $R_{\mathrm{eff}}(A, B) = R_{\mathrm{eff}}(B, A)$. If $A \subset A'$, $B \subset B'$ with $A' \cap B' = \emptyset$, then $R_{\mathrm{eff}}(A', B') \leq R_{\mathrm{eff}}(A, B)$.

Take a bond $e = \{x, y\}$, $x \sim y$ in a weighted graph (X, μ). We say *cutting the bond e* when we take the conductance μ_{xy} to be 0, and we say *shorting the bond e* when we identify $x = y$ and take the conductance μ_{xy} to be ∞. Clearly, shorting decreases the effective resistance (*shorting law*), and cutting increases the effective resistance (*cutting law*).

The following proposition (*Dirichlet's principle*) shows that among feasible potentials whose voltage is 1 on A and 0 on B, it is a harmonic function on $(A \cup B)^c$ that minimizes the energy.

Proposition 2.2.3. *Assume $R_{\text{eff}}(A, B) \neq 0$.*

(i) The right hand side of (2.16) is attained by a unique minimizer φ.
(ii) φ in (i) is a solution of the following Dirichlet problem

$$\begin{cases} \mathcal{L}\varphi(x) = 0, \ \forall x \in X \setminus (A \cup B), \\ \varphi|_A = 1, \ \varphi|_B = 0. \end{cases} \tag{2.17}$$

Proof. (i) We fix a base point $x_0 \in B$ and recall that H^2 is a Hilbert space with $\|f\|_{H^2} = \mathcal{E}(f, f) + f(x_0)^2$ (Lemma 2.1.2 (ii)). Since $\mathcal{V} := \{f \in H^2 : f|_A = 1, f|_B = 0\}$ is a non-void (because $R_{\text{eff}}(A, B) \neq 0$) closed convex subset of H^2, a general theorem shows that $\mathcal{V}$ has a unique minimizer for $\| \cdot \|_{H^2}$ (which is equal to $\mathcal{E}(\cdot, \cdot)$ on $\mathcal{V}$).

(ii) Let g be a function on X whose support is finite and is contained in $X \setminus (A \cup B)$. Then, for any $\lambda \in \mathbb{R}$, $\varphi + \lambda g \in \mathcal{V}$, so $\mathcal{E}(\varphi + \lambda g, \varphi + \lambda g) \geq \mathcal{E}(\varphi, \varphi)$. Thus $\mathcal{E}(\varphi, g) = 0$. Applying Lemma 2.1.5 (ii), we have $(\mathcal{L}\varphi, g) = 0$. For each $x \in X \setminus (A \cup B)$, by choosing $g(z) = \delta_x(z)$, we obtain $\mathcal{L}\varphi(x)\mu_x = 0$. $\square$

As we mentioned in Remark 2.2.2 (ii), we do not have uniqueness of the Dirichlet problem in general. So in the following of this section, we will assume that A^c is finite in order to guarantee uniqueness of the Dirichlet problem.

Remark 2.2.4. There is a dual characterization of resistance using flows of the network. It is called *Thompson's principle* (see for example, [20, 93]).

The next theorem gives a probabilistic interpretation of the effective resistance.

Theorem 2.2.5. *If A^c is finite, then for each $x_0 \in A^c$,*

$$R_{\text{eff}}(x_0, A)^{-1} = \mu_{x_0} \mathbb{P}^{x_0}(\sigma_A < \sigma_{x_0}^+). \tag{2.18}$$

Proof. Let $v(x) = \mathbb{P}^x(\sigma_A < \sigma_{x_0})$. Then, by Proposition 2.2.1, v is the unique solution of Dirichlet problem with $v(x_0) = 0$, $v|_A = 1$. By Proposition 2.2.3 and Lemma 2.1.5 (noting that $1 - v \in \mathbb{L}^2$),

$$R_{\text{eff}}(x_0, A)^{-1} = \mathcal{E}(v, v) = \mathcal{E}(-v, 1 - v) = (\mathcal{L}v, 1 - v)$$

$$= \mathcal{L}v(x_0)\mu_{x_0} = \mathbb{E}^{x_0}[v(Y_1)]\mu_{x_0}.$$

By definition of v, one can see $\mathbb{E}^{x_0}[v(Y_1)] = \mathbb{P}^{x_0}(\sigma_A < \sigma_{x_0}^+)$ so the result follows.

$\square$

Similarly, if A^c is finite one can prove

$$R_{\text{eff}}(B, A)^{-1} = \sum_{x \in B} \mu_x \mathbb{P}^x (\sigma_A < \sigma_B^+).$$

Note that by Ohm's law, the right hand side of (2.18) is the current flowing from x_0 to A.

The following lemma is useful and will be used later in Proposition 4.4.3.

Lemma 2.2.6. *Let* $A, B \subset X$ *and assume that both* A^c, B^c *are finite. Then the following holds for all* $x \notin A \cup B$.

$$(R_{\text{eff}}(x, A \cup B))(R_{\text{eff}}(x, A)^{-1} - R_{\text{eff}}(x, B)^{-1}) \leq \mathbb{P}^x (\sigma_A < \sigma_B) \leq \frac{R_{\text{eff}}(x, A \cup B)}{R_{\text{eff}}(x, A)}.$$

Proof. Using the strong Markov property, we have

$$\mathbb{P}^x (\sigma_A < \sigma_B) = \mathbb{P}^x (\sigma_A < \sigma_B, \sigma_{A \cup B} < \sigma_x^+) + \mathbb{P}^x (\sigma_A < \sigma_B, \sigma_{A \cup B} > \sigma_x^+)$$
$$= \mathbb{P}^x (\sigma_A < \sigma_B, \sigma_{A \cup B} < \sigma_x^+) + \mathbb{P}^x (\sigma_{A \cup B} > \sigma_x^+) \mathbb{P}^x (\sigma_A < \sigma_B).$$

So

$$\mathbb{P}^x (\sigma_A < \sigma_B) = \frac{\mathbb{P}^x (\sigma_A < \sigma_B, \sigma_{A \cup B} < \sigma_x^+)}{\mathbb{P}^x (\sigma_{A \cup B} < \sigma_x^+)} \leq \frac{\mathbb{P}^x (\sigma_A < \sigma_x^+)}{\mathbb{P}^x (\sigma_{A \cup B} < \sigma_x^+)}.$$

Using (2.18), the upper bound is obtained. For the lower bound,

$$\mathbb{P}^x (\sigma_A < \sigma_B, \sigma_{A \cup B} < \sigma_x^+) \geq \mathbb{P}^x (\sigma_A < \sigma_x^+ < \sigma_B)$$
$$\geq \mathbb{P}^x (\sigma_A < \sigma_x^+) - \mathbb{P}^x (\sigma_B < \sigma_x^+),$$

so using (2.18) again, the proof is complete.   □

As we see in the proof, we only need to assume that A^c is finite for the upper bound.

Now let (X, μ) be an infinite weighted graph. Let $\{A_n\}_{n=1}^{\infty}$ be a family of finite sets such that $A_n \subset A_{n+1}$ for $n \in \mathbb{N}$ and $\cup_{n \geq 1} A_n = X$. Let $x_0 \in A_1$. By the shorting law, $R_{\text{eff}}(x_0, A_n^c) \leq R_{\text{eff}}(x_0, A_{n+1}^c)$, so the following limit exists.

$$R_{\text{eff}}(x_0) := \lim_{n \to \infty} R_{\text{eff}}(x_0, A_n^c). \tag{2.19}$$

Further, the limit $R_{\text{eff}}(x_0)$ is independent of the choice of the sequence $\{A_n\}$ mentioned above. (Indeed, if $\{B_n\}$ is another such family, then for each n there exists N_n such that $A_n \subset B_{N_n}$, so $\lim_{n \to \infty} R_{\text{eff}}(x_0, A_n^c) \leq \lim_{n \to \infty} R_{\text{eff}}(x_0, B_n^c)$. By changing the role of A_n and B_n, we have the opposite inequality.)

Theorem 2.2.7. *Let (X, μ) be an infinite weighted graph. For each $x \in X$, the following holds*

$$\mathbb{P}^x(\sigma_x^+ = \infty) = (\mu_x R_{\mathrm{eff}}(x))^{-1}.$$

Proof. By Theorem 2.2.5, we have

$$\mathbb{P}^x(\sigma_{A_n^c} < \sigma_x^+) = (\mu_x R_{\mathrm{eff}}(x, A_n^c))^{-1}.$$

Taking $n \to \infty$ and using (2.19), we have the desired equality. □

Definition 2.2.8. We say a Markov chain is recurrent at $x \in X$ if $\mathbb{P}^x(\sigma_x^+ = \infty) = 0$. We say a Markov chain is transient at $x \in X$ if $\mathbb{P}^x(\sigma_x^+ = \infty) > 0$.

The following is well-known for irreducible Markov chains (so in particular it holds for Markov chains corresponding to weighted graphs). See for example [188].

Proposition 2.2.9. *(1) $\{Y_n\}_n$ is recurrent at $x \in X$ if and only if $m := \sum_{n=0}^{\infty} \mathbb{P}^x(Y_n = x) = \infty$. Further, $m^{-1} = \mathbb{P}^x(\sigma_x^+ = \infty)$.*
(2) If $\{Y_n\}_n$ is recurrent (resp. transient) at some $x \in X$, then it is recurrent (resp. transient) for all $x \in X$.
(3) $\{Y_n\}_n$ is recurrent if and only if $\mathbb{P}^x(\{Y$ hits y infinitely often$\}) = 1$ for all $x, y \in X$. $\{Y_n\}_n$ is transient if and only if $\mathbb{P}^x(\{Y$ hits y finitely often$\}) = 1$ for all $x, y \in X$.

From Theorem 2.2.7 and Proposition 2.2.9, we have the following.

$\{Y_n\}$ is transient (resp. recurrent)

$$\Leftrightarrow R_{\mathrm{eff}}(x) < \infty \text{ (resp. } R_{\mathrm{eff}}(x) = \infty), \exists x \in X \qquad (2.20)$$

$$\Leftrightarrow R_{\mathrm{eff}}(x) < \infty \text{ (resp. } R_{\mathrm{eff}}(x) = \infty), \forall x \in X.$$

Example 2.2.10. Consider $\mathbb{Z}^2$ with weight 1 on each nearest neighbor bond. Let $\partial B_n = \{(x, y) \in \mathbb{Z}^2 : \text{either } |x| \text{ or } |y| \text{ is } n\}$. By shorting ∂B_n for all $n \in \mathbb{N}$, one can obtain

$$R_{\mathrm{eff}}(0) \geq \sum_{n=0}^{\infty} \frac{1}{4(2n+1)} = \infty.$$

So the simple random walk on $\mathbb{Z}^2$ is recurrent.

Let us recall the following fact.

Theorem 2.2.11 (Pólya 1921). *Simple random walk on $\mathbb{Z}^d$ is recurrent if $d = 1, 2$ and transient if $d \geq 3$.*

The combinatorial proof of this theorem is well-known. For example, for $d = 1$, by counting the total number of paths of length $2n$ that moves both right and left n times,

$$\mathbb{P}^0(Y_{2n} = 0) = 2^{-2n} \binom{2n}{n} = \frac{(2n)!}{2^{2n} n! n!} \sim (\pi n)^{-1/2},$$

where Stirling's formula is used in the end. Thus

$$m = \sum_{n=0}^{\infty} \mathbb{P}^0(Y_n = 0) \sim \sum_{n=1}^{\infty} (\pi n)^{-1/2} + 1 = \infty,$$

so $\{Y_n\}$ is recurrent.

This argument is not robust. For example, if one changes the weight on $\mathbb{Z}^d$ so that $c_1 \leq \mu_{xy} \leq c_2$ for $x \sim y$, one cannot apply the argument at all. The advantage of the characterization of transience/recurrence using the effective resistance is that one can make a robust argument. Indeed, by (2.20) we can see that transience/recurrence is stable under bounded perturbation. This is because, if $c_1 \mu'_{xy} \leq \mu_{xy} \leq c_2 \mu'_{xy}$ for all $x, y \in X$, then $c_1 R_{\mathrm{eff}}(x) \leq R'_{\mathrm{eff}}(x) \leq c_2 R_{\mathrm{eff}}(x)$. We can further prove that transience/recurrence is stable under rough isometry.

Finally in this section, we will give more equivalence condition for the transience and discuss some decomposition of H^2. Let H_0^2 be the closure of $C_0(X)$ in H^2, where $C_0(X)$ is the space of compactly supported functions on X. For a finite set $B \subset X$, define the capacity of B by

$$\mathrm{Cap}\,(B) = \inf\{\mathcal{E}(f, f) : f \in H_0^2, f|_B = 1\}.$$

We first give a lemma.

Lemma 2.2.12. *If a sequence of non-negative functions $v_n \in H^2$, $n \in \mathbb{N}$ satisfies $\lim_{n \to \infty} v_n(x) = \infty$ for all $x \in X$ and $\lim_{n \to \infty} \mathcal{E}(v_n, v_n) = 0$, then*

$$\lim_{n \to \infty} \|u - (u \wedge v_n)\|_{H^2} = 0, \qquad \forall u \in H^2, u \geq 0.$$

Proof. Let $u_n = u \wedge v_n$ and define $U_n = \{x \in X : u(x) > v_n(x)\}$. By the assumption, for each $N \in \mathbb{N}$, there exists $N_0 = N_0(N)$ such that $U_n \subset B(0, N)^c$ for all $n \geq N_0$. For $A \subset X$, denote $\mathcal{E}_A(u) = \frac{1}{2} \sum_{x,y \in A} (u(x) - u(y))^2 \mu_{xy}$. Since $\mathcal{E}_{U_n^c}(u - u_n) = 0$, we have

$$\mathcal{E}(u - u_n, u - u_n) \leq 2 \cdot \frac{1}{2} \sum_{x \in U_n} \sum_{y : y \sim x} \Big(u(x) - u_n(x) - (u(y) - u_n(y)) \Big)^2 \mu_{xy}$$

$$\leq 2\mathcal{E}_{B(0,N-1)^c}(u - u_n) \leq 4 \Big(\mathcal{E}_{B(0,N-1)^c}(u) + \mathcal{E}_{B(0,N-1)^c}(u_n) \Big) \qquad (2.21)$$

for all $n \geq N_0$. As $u_n = (u + v_n - |u - v_n|)/2$, we have

$$\mathcal{E}_{B(0,N-1)^c}(u_n) \leq c_1 \Big(\mathcal{E}_{B(0,N-1)^c}(u) + \mathcal{E}_{B(0,N-1)^c}(v_n) + \mathcal{E}_{B(0,N-1)^c}(|u - v_n|) \Big)$$

$$\leq c_2 \Big(\mathcal{E}_{B(0,N-1)^c}(u) + \mathcal{E}_{B(0,N-1)^c}(v_n) \Big).$$

Thus, together with (2.21), we have

$$\mathcal{E}(u - u_n, u - u_n) \leq c_3 \left(\mathcal{E}_{B(0,N-1)^c}(u) + \mathcal{E}_{B(0,N-1)^c}(v_n) \right)$$

$$\leq c_3 \left(\mathcal{E}_{B(0,N-1)^c}(u) + \mathcal{E}(v_n, v_n) \right).$$

Since $u \in H^2$, $\mathcal{E}_{B(0,N-1)^c}(u) \to 0$ as $N \to \infty$ and by the assumption, $\mathcal{E}(v_n, v_n) \to 0$ as $n \to \infty$. So we obtain $\mathcal{E}(u - u_n, u - u_n) \to 0$ as $n \to \infty$. By the assumption, it is clear that $u - u_n \to 0$ pointwise, so we obtain $\|u - u_n\|_{H^2} \to 0$. □

We say that a quadratic form $(\mathcal{E}, \mathcal{F})$ is Markovian if $u \in \mathcal{F}$ and $v = (0 \vee u) \wedge 1$, then $v \in \mathcal{F}$ and $\mathcal{E}(v, v) \leq \mathcal{E}(u, u)$. It is easy to see that quadratic forms determined by weighted graphs are Markovian.

Proposition 2.2.13. *The following are equivalent.*

 (i) *The Markov chain corresponding to (X, μ) is transient.*
 (ii) $1 \notin H_0^2$
(iii) *Cap$(\{x\}) > 0$ for some $x \in X$.*
(iii)' *Cap$(\{x\}) > 0$ for all $x \in X$.*
 (iv) $H_0^2 \neq H^2$
 (v) *There exists a non-negative super-harmonic function which is not a constant function.*
 (vi) *For each $x \in X$, there exists $c_1(x) > 0$ such that*

$$|f(x)|^2 \leq c_1(x) \mathcal{E}(f, f) \qquad \forall f \in C_0(X). \tag{2.22}$$

Proof. For fixed $x \in X$, define $\varphi(z) = \mathbb{P}^z(\sigma_x < \infty)$. We first show the following: $\varphi \in H_0^2$ and

$$\mathcal{E}(\varphi, \varphi) = (-\mathcal{L}\varphi, 1_{\{x\}}) = R_{\text{eff}}(x)^{-1} = \text{Cap}(\{x\}). \tag{2.23}$$

Indeed, let $\{A_n\}_{n=1}^{\infty}$ be a family of finite sets such that $A_n \subset A_{n+1}$ for $n \in \mathbb{N}$, $x \in A_1$, and $\cup_{n \geq 1} A_n = X$. Then $R_{\text{eff}}(x, A_n^c)^{-1} \downarrow R_{\text{eff}}(x)^{-1}$. Let $\varphi_n(z) = \mathbb{P}^z(\sigma_x < \tau_{A_n})$. Using Lemma 2.1.5 (ii), and noting $\varphi_n \in C_0(X)$, we have, for $m \leq n$,

$$\mathcal{E}(\varphi_m, \varphi_n) = (\varphi_m, -\mathcal{L}\varphi_n) = (1_{\{x\}}, -\mathcal{L}\varphi_n) = \mathcal{E}(\varphi_n, \varphi_n) = R_{\text{eff}}(x, A_n^c)^{-1}. \tag{2.24}$$

This implies

$$\mathcal{E}(\varphi_m - \varphi_n, \varphi_m - \varphi_n) = R_{\text{eff}}(x, A_m^c)^{-1} - R_{\text{eff}}(x, A_n^c)^{-1}.$$

Hence $\{\varphi_m\}$ is a $\mathcal{E}$-Cauchy sequence. Noting that $\varphi_n \to \varphi$ pointwise, we see that $\varphi_n \to \varphi$ in H^2 as well and $\varphi \in H_0^2$. Taking $n = m$ and $n \to \infty$ in (2.24), we obtain (2.23) except the last equality. To prove the last equality of (2.23), take any $f \in H_0^2$

with $f(x) = 1$. Then $g := f - \varphi \in H_0^2$ and $g(x) = 0$. Let $g_n \in C_0(X)$ with $g_n \to g$ in H_0^2. Then, by Lemma 2.1.5 (ii), $\mathcal{E}(\varphi, g_n) = (-\mathcal{L}\varphi, g_n)$. Noting that φ is harmonic except at x, we see that $\mathcal{L}\varphi \in C_0(X)$. So, letting $n \to \infty$, we have

$$\mathcal{E}(\varphi, g) = (-\mathcal{L}\varphi, g) = -\mathcal{L}\varphi(x)g(x)\mu_x = 0.$$

Thus,

$$\mathcal{E}(f, f) = \mathcal{E}(\varphi + g, \varphi + g) = \mathcal{E}(\varphi, \varphi) + \mathcal{E}(g, g) \geq \mathcal{E}(\varphi, \varphi),$$

which means that φ is the unique minimizer in the definition of Cap $(\{x\})$. So the last equality of (2.23) is obtained.

Given (2.23), we now prove the equivalence.

$(i) \implies (iii)'$: This is a direct consequence of (2.20) and (2.23).

$(iii) \iff (ii) \iff (iii)'$: This is easy. Indeed, Cap $(\{x\}) = 0$ if and only if there is $f \in H_0^2$ with $f(x) = 1$ and $\mathcal{E}(f, f) = 0$, that is f is identically 1.

$(iii)' \implies (vi)$: Let $f \in C_0(X) \subset H_0^2$ with $f(x) \neq 0$, and define $g = f/f(x)$. Then

$$\text{Cap}\,(\{x\}) \leq \mathcal{E}(g, g) = \mathcal{E}(f, f)/f(x)^2.$$

So, letting $c_1(x) = 1/\text{Cap}\,(\{x\}) > 0$, we obtain (vi).

$(vi) \implies (i)$: As before, let $\varphi_n(z) = \mathbb{P}^z(\sigma_x < \tau_{A_n})$. Then by (2.22), $\mathcal{E}(\varphi_n, \varphi_n) \geq c_1(x)^{-1}$. So, using the fact $\varphi_n \to \varphi$ in H^2 and (2.23), $R_{\text{eff}}(x)^{-1} = \mathcal{E}(\varphi, \varphi) = \lim_n \mathcal{E}(\varphi_n, \varphi_n) \geq c_1(x)^{-1}$. This means the transience by (2.20).

$(ii) \iff (iv)$: $(ii) \implies (iv)$ is clear since $1 \in H^2$, so we will prove the opposite direction. Suppose $1 \in H_0^2$. Then there exists $\{f_n\}_n \subset C_0(X)$ such that $\|1 - f_n\|_{H^2} < n^{-2}$. Since $\mathcal{E}$ is Markovian, we have $\|1 - f_n\|_{H^2} \geq \|1 - (f_n \vee 0) \wedge 1\|_{H^2}$, so without loss of generality we may assume $f_n \geq 0$. Let $v_n = nf_n \geq 0$. Then $\lim_n v_n(x) = \infty$ for all $x \in X$ and $\mathcal{E}(v_n, v_n) = n^2 \mathcal{E}(f_n, f_n) \leq n^{-2} \to 0$ so by Lemma 2.2.12, $\|u - (u \wedge v_n)\|_{H^2} \to 0$ for all $u \in H^2$ with $u \geq 0$. Since $u \wedge v_n \in C_0(X)$, this implies $u \in H_0^2$. For general $u \in H^2$, we can decompose it into $u_+ - u_-$ where $u_+, u_- \geq 0$ are in H^2. So applying the above, we have $u_+, u_- \in H_0^2$ and conclude $u \in H_0^2$.

$(i) \implies (v)$: If the corresponding Markov chain is transient, then $\psi(z) = \mathbb{P}^z(\sigma_x^+ < \infty)$ is the non-constant super-harmonic function.

$(i) \impliedby (v)$: Suppose the corresponding Markov chain $\{Y_n\}_n$ is recurrent. For a super-harmonic function $\psi \geq 0$, $M_n = \psi(Y_n) \geq 0$ is a supermartingale, so it converges $\mathbb{P}^x$-a.s. Let M_∞ be the limiting random variable. Since the set $\{n \in \mathbb{N} : Y_n = y\}$ is unbounded $\mathbb{P}^x$-a.s. for all $y \in X$ (due to the recurrence), we have $\mathbb{P}^x(\psi(y) = M_\infty) = 1$ for all $y \in X$, so ψ is constant. □

Remark 2.2.14. $(v) \implies (i)$ implies that if the Markov chain corresponding to (X, μ) is recurrent, then it has the strong Liouville property.

For A, B which are subspaces of H^2, we write $A \oplus B = \{f + g : f \in A, g \in B\}$ if $\mathcal{E}(f, g) = 0$ for all $f \in A$ and $g \in B$.

As we see above, the Markov chain corresponding to (X, μ) is recurrent if and only if $H^2 = H_0^2$. When the Markov chain is transient, we have the following decomposition of H^2, which is called the Royden decomposition (see [204, Theorem 3.69]).

Proposition 2.2.15. *Suppose that the Markov chain corresponding to (X, μ) is transient. Then*

$$H^2 = \mathcal{H} \oplus H_0^2,$$

where $\mathcal{H} := \{h \in H^2 : h \text{ is a harmonic functions on } X\}$. Further the decomposition is unique.

Proof. For each $f \in H^2$, let $a_f = \inf_{h \in H_0^2} \mathcal{E}(f - h, f - h)$. Then, similarly to the proof of Proposition 2.2.3, we can show that there is a unique minimizer $v_f \in H_0^2$ such that $a_f = \mathcal{E}(f - v_f, f - v_f)$, $\mathcal{E}(f - v_f, g) = 0$ for all $g \in H_0^2$, and in particular $f - v_f$ is harmonic on X. For the uniqueness of the decomposition, suppose $f = u + v = u' + v'$ where $u, u' \in \mathcal{H}$ and $v, v' \in H_0^2$. Then, $w := u - u' = v' - v \in \mathcal{H} \cap H_0^2$, so $\mathcal{E}(w, w) = 0$, which implies w is constant. Since $w \in H_0^2$ and the Markov chain is transient, by Proposition 2.2.13 we have $w \equiv 0$. $\square$

2.3   Trace of Weighted Graphs

Finally in this chapter, we briefly mention the trace of weighted graphs, which will be used in Chaps. 4 and 8. Note that there is a general theory on traces for Dirichlet forms (see [74, 108]). Also note that a trace to infinite subset of X may not satisfy locally finiteness, but one can consider quadratic forms on them similarly.

Proposition 2.3.1 (Trace of the Weighted Graph). *Let $V \subset X$ be a non-void set such that $\mathbb{P}^x(\sigma_V < \infty) = 1$ for all $x \in X$ and let $f \in H^2(V) := \{u|_V : u \in H^2\}$. Then there exists a unique $u \in H^2$ which attains the following infimum:*

$$\inf\{\mathcal{E}(v, v) : v \in H^2, v|_V = f\}. \tag{2.25}$$

Moreover, the map $f \mapsto u =: H_V f$ is a linear map and there exist weights $\{\hat{\mu}_{xy}\}_{x,y \in V}$ such that the corresponding quadratic form $\hat{\mathcal{E}}_V(\cdot, \cdot)$ satisfies the following:

$$\hat{\mathcal{E}}_V(f, f) = \mathcal{E}(H_V f, H_V f) \qquad \forall f \in H^2(V).$$

Proof. The proof here is inspired by [153, Sect. 8].

The fact that there exists a unique $u \in H^2$ that attains the infimum of (2.25) can be proved similarly to Proposition 2.2.3 (i). So the map $H_V : H^2(V) \to H^2$ where $f \mapsto H_V f$ is well-defined. Let $\mathcal{H}_V := \{u \in H^2 : \mathcal{E}(u, v) = 0,$ for all $v \in H^2$ such that $v|_V = 0\}$; a space of harmonic functions on $X \setminus V$. We claim the following:

$$\mathcal{H}_V = H_V(H^2(V)) \quad \text{and} \quad R_V : \mathcal{H}_V \to H^2(V), \tag{2.26}$$

where $R_V u = u|_V$ is an inverse operator of H_V. Once this is proved, then we have the linearity of H_V and furthermore we have

$$H^2 = \mathcal{H}_V \oplus \{v \in H^2 : v|_V = 0\}.$$

So let us prove (2.26). If $f \in H^2(V)$ and $u = H_V f$, then for any $v \in H^2$ with $v|_V = f$, we have

$$\mathcal{E}(\lambda(v - u) + u, \lambda(v - u) + u) \geq \mathcal{E}(u, u), \qquad \forall \lambda \in \mathbb{R},$$

because u attains the infimum in (2.25). This implies $\mathcal{E}(v - u, u) = 0$, namely $u \in \mathcal{H}_V$. Clearly $u|_V = f$, so we obtain $\mathcal{H}_V \supset H_V(H^2(V))$ and $R_V \circ H_V$ is an identity map. Next, if $u \in \mathcal{H}_V$ and $u|_V = f \in H^2(V)$, then for any $v \in H^2$ with $v|_V = f$, we have

$$\mathcal{E}(v, v) = \mathcal{E}(v - u + u, v - u + u) = \mathcal{E}(v - u, v - u) + \mathcal{E}(u, u) \geq \mathcal{E}(u, u)$$

because $\mathcal{E}(v - u, u) = 0$ (since $u \in \mathcal{H}_V$). This implies $u = H_V f$, since the infimum in (2.25) is attained uniquely by $H_V f$. So we obtain $\mathcal{H}_V \subset H_V(H^2(V))$ and $H_V \circ R_V$ is an identity map.

Set $\hat{\mathcal{E}}(f, f) = \mathcal{E}(H_V f, H_V f)$. Clearly, $\hat{\mathcal{E}}$ is a non-negative definite symmetric bilinear form and $\hat{\mathcal{E}}(f, f) = 0$ if any only if f is a constant function. So, there exists $\{a_{xy}\}_{x,y \in V}$ with $a_{xy} = a_{yx}$ such that $\hat{\mathcal{E}}(f, f) = \frac{1}{2} \sum_{x,y \in V} a_{xy}(f(x) - f(y))^2$.

Next, we show that $\hat{\mathcal{E}}$ is Markovian. Indeed, writing $\bar{u} = (0 \vee u) \wedge 1$ for a function u, since $\overline{H_V u}|_V = \bar{u}$, we have

$$\hat{\mathcal{E}}(\bar{u}, \bar{u}) = \mathcal{E}(H_V \bar{u}, H_V \bar{u}) \leq \mathcal{E}(\overline{H_V u}, \overline{H_V u}) \leq \mathcal{E}(H_V u, H_V u) = \hat{\mathcal{E}}(u, u),$$

for all $u \in H^2(V)$, where the fact that $\mathcal{E}$ is Markovian is used in the second inequality. Now take $p, q \in V$ with $p \neq q$ arbitrary, and consider a function h such that $h(p) = 1, h(q) = -\alpha < 0$ and $h(z) = 0$ for $z \in V \setminus \{p, q\}$. Then, there exist c_1, c_2 such that

$$\hat{\mathcal{E}}(h, h) = a_{pq}(h(p) - h(q))^2 + c_1 h(p)^2 + c_2 h(q)^2 = a_{pq}(1 + \alpha)^2 + c_1 + c_2 \alpha^2$$
$$\geq \hat{\mathcal{E}}(\bar{h}, \bar{h}) = a_{pq}(\bar{h}(p) - \bar{h}(q))^2 + c_1 \bar{h}(p)^2 + c_2 \bar{h}(q)^2 = a_{pq} + c_1.$$

So $(a_{pq} + c_2)\alpha^2 + 2a_{pq}\alpha \geq 0$. Since this holds for all $\alpha > 0$, we have $a_{pq} \geq 0$. Putting $\hat{\mu}_{pq} = a_{pq}$ for each $p, q \in V$ with $p \neq q$, we have $\hat{\mathcal{E}}_V = \hat{\mathcal{E}}$, that is $\hat{\mathcal{E}}$ is associated with the weighted graph $(V, \hat{\mu})$.   □

We call the induced weights $\{\hat{\mu}_{xy}\}_{x,y \in V}$ as the *trace of* $\{\mu_{xy}\}_{x,y \in X}$ *to* V. From this proposition, we see that for $x, y \in V$, $R_{\text{eff}}(x, y) = R^V_{\text{eff}}(x, y)$ where $R^V_{\text{eff}}(\cdot, \cdot)$ is the effective resistance for $(V, \hat{\mu})$.

Chapter 3
Heat Kernel Estimates: General Theory

In this chapter, we will consider various inequalities which imply (or which are equivalent to) the Nash-type heat kernel upper bound, i.e. $p_t(x, y) \leq c_1 t^{-\theta/2}$ for some $\theta > 0$. We will also discuss Poincaré inequalities and their relations to heat kernel estimates in Sect. 3.3. We will prefer to discuss them under a general framework including weighted graphs. However, some arguments here are rather sketchy to apply for the general framework. (Readers may only consider weighted graphs throughout the chapter.) Sections 3.1 and 3.2 are strongly motivated by Coulhon's survey paper [75].

Let X be a locally compact separable metric space and μ be a Radon measure on X such that $\mu(B) > 0$ for any non-void ball. $(\mathcal{E}, \mathcal{F})^1$ is called a Dirichlet form on $\mathbb{L}^2(X, \mu)$ if it is a symmetric closed bilinear Markovian form on $\mathbb{L}^2$. It is well-known that given a Dirichlet form, there is a corresponding symmetric strongly continuous Markovian semigroup $\{P_t\}_{t \geq 0}$ on $\mathbb{L}^2(X, \mu)$ (see [108, Sect. 1.3, 1.4]). Here the Markovian property of the semigroup means if $u \in \mathbb{L}^2$ satisfies $0 \leq u \leq 1$ μ-a.s., then $0 \leq P_t u \leq 1$ μ-a.s. for all $t \geq 0$. We denote the corresponding non-negative definite $\mathbb{L}^2$-generator by $-\mathcal{L}$.

We denote the inner product of $\mathbb{L}^2$ by $(\cdot, \cdot)$ and for $p \geq 1$ denote $\|f\|_p$ for the $\mathbb{L}^p$-norm of $f \in \mathbb{L}^2(X, \mu)$. For each $\alpha > 0$, define

$$\mathcal{E}_\alpha(\cdot, \cdot) = \mathcal{E}(\cdot, \cdot) + \alpha(\cdot, \cdot).$$

$(\mathcal{E}_1, \mathcal{F})$ is then a Hilbert space.

---

[1] $\mathcal{F}$ is a domain of the Dirichlet form. For example, when $\mathcal{E}(f, f) = \frac{1}{2} \int_{\mathbb{R}^d} |\nabla f|^2 dx$, then $\mathcal{F} = W^{1,2}(\mathbb{R}^d)$, the classical Sobolev space. When we consider weighted graphs, we may take $\mathcal{F} = \mathbb{L}^2(X, \mu)$ (or $\mathcal{F} = H^2$).

T. Kumagai, *Random Walks on Disordered Media and their Scaling Limits*, Lecture Notes in Mathematics 2101, DOI 10.1007/978-3-319-03152-1_3,
© Springer International Publishing Switzerland 2014

3.1   The Nash Inequality

We first give a preliminary lemma.

Lemma 3.1.1. *(i)* $\|P_t f\|_1 \le \|f\|_1$ *for all* $f \in \mathbb{L}^1 \cap \mathbb{L}^2$.
(ii) For $f \in \mathbb{L}^2$, *define* $u(t) = (P_t f, P_t f)$. *Then* $u'(t) = -2\mathcal{E}(P_t f, P_t f)$.
(iii) For $f \in \mathcal{F}$ *and* $t \ge 0$, $\exp(-\mathcal{E}(f, f)t/\|f\|_2^2) \le \|P_t f\|_2/\|f\|_2$.

Proof. (i) We first show that if $0 \le f \in \mathbb{L}^2$, then $0 \le P_t f$. Indeed, if we let
$f_n = f \wedge n$, then $f_n \to f$ in $\mathbb{L}^2$. Since $0 \le f_n \le n$, the Markovian property
of $\{P_t\}$ implies that $0 \le P_t f_n \le n$. Taking $n \to \infty$, we obtain $0 \le P_t f$. So,
for $f \in \mathbb{L}^2$ we have $P_t|f| \ge |P_t f|$, since $-|f| \le f \le |f|$. Using this and
the Markovian property, we have for all $f \in \mathbb{L}^2 \cap \mathbb{L}^1$ and all Borel set $A \subset X$,

$$(|P_t f|, 1_A) \le (P_t|f|, 1_A) = (|f|, P_t 1_A) \le \|f\|_1.$$

Hence we have $P_t f \in \mathbb{L}^1$ and $\|P_t f\|_1 \le \|f\|_1$.
(ii) Since $P_t f \in Dom(\mathcal{L})$, we have

$$
\begin{aligned}
\frac{u(t+h) - u(t)}{h} &= \frac{1}{h}(P_{t+h} f + P_t f, P_{t+h} f - P_t f) \\
&= (P_{t+h} f + P_t f, \frac{(P_h - I)P_t f}{h}) \\
&\xrightarrow{h \downarrow 0} 2(P_t f, \mathcal{L} P_t f) = -2\mathcal{E}(P_t f, P_t f).
\end{aligned}
$$

Hence $u'(t) = -2\mathcal{E}(P_t f, P_t f)$.
(iii) We will prove the inequality for $f \in Dom(\mathcal{L})$; then one can obtain the
result for $f \in \mathcal{F}$ by approximation. Let $-\mathcal{L} = \int_0^\infty \lambda dE_\lambda$ be the spectral
decomposition of $-\mathcal{L}$. Then $P_t = e^{\mathcal{L}t} = \int_0^\infty e^{-\lambda t} dE_\lambda$ and $\|f\|_2^2 = \int_0^\infty (dE_\lambda f, f)$. Since $\lambda \mapsto e^{-2\lambda t}$ is convex, by Jensen's inequality,

$$\exp\left(-2 \int_0^\infty \lambda t \frac{(dE_\lambda f, f)}{\|f\|_2^2}\right) \le \int_0^\infty e^{-2\lambda t} \frac{(dE_\lambda f, f)}{\|f\|_2^2} = \frac{(P_{2t} f, f)}{\|f\|_2^2} = \frac{\|P_t f\|_2^2}{\|f\|_2^2}.$$

Taking the square root in each term, we obtain the desired inequality.   □

Remark 3.1.2. An alternative proof of (iii) is to use the logarithmic convexity of
$\|P_t f\|_2^2$. Indeed,

$$\|P_{(t+s)/2} f\|_2^2 = (P_{t+s} f, f) = (P_t f, P_s f) \le \|P_t f\|_2 \|P_s f\|_2, \quad \forall s, t > 0$$

so $\|P_t f\|_2^2$ is logarithmic convex. Thus,

$$t \mapsto \frac{d}{dt} \log \|P_t f\|_2^2 = \frac{\frac{d}{dt}(\|P_t f\|_2^2)}{\|P_t f\|_2^2} = -\frac{2\mathcal{E}(P_t f, P_t f)}{\|P_t f\|_2^2} \tag{3.1}$$

is non-decreasing. (The last equality is due to Lemma 3.1.1 (ii).) The right hand side of (3.1) is $-\frac{2\mathcal{E}(f,f)}{\|f\|_2^2}$ when $t = 0$, so integrating (3.1) over $[0,t]$, we have

$$\log \frac{\|P_t f\|_2^2}{\|f\|_2^2} = \int_0^t \frac{d}{ds} \log \|P_s f\|_2^2 ds \geq -\frac{2t\mathcal{E}(f,f)}{\|f\|_2^2}.$$

The following is easy to see. (Note that we only need the first assertion. We refer the terminology of Dirichlet forms to [74, 108].)

Lemma 3.1.3. *Let $(\mathcal{E}, \mathcal{F})$ be a symmetric closed bilinear form on $\mathbb{L}^2(X, \mu)$, and let $\{P_t\}_{t\geq 0}$, $-\mathcal{L}$ be the corresponding semigroup and the self-adjoint operator respectively. Then, for each $\delta > 0$, $(\mathcal{E}_\delta, \mathcal{F})$ is also a symmetric closed bilinear form and the corresponding semigroup and the self-adjoint operator are $\{e^{-\delta t}P_t\}_{t\geq 0}$, $\delta I - \mathcal{L}$, respectively. Further, if $(\mathcal{E}, \mathcal{F})$ is the regular Dirichlet form on $\mathbb{L}^2(X, \mu)$ and the corresponding Hunt process is $\{Y_t\}_{t\geq 0}$, then $(\mathcal{E}_\delta, \mathcal{F})$ is also the regular Dirichlet form and the corresponding hunt process is $\{Y_{t\wedge\zeta}\}_{t\geq 0}$ where ζ is the independent exponential random variable with parameter δ. (ζ is the killing time; i.e. the process goes to the cemetery point at ζ.)*

The next theorem is by Carlen-Kusuoka-Stroock [69], where the original idea of the proof of $(i) \Rightarrow (ii)$ is due to Nash [187].

Theorem 3.1.4 (The Nash Inequality, [69]). *The following are equivalent for any $\delta \geq 0$.*

(i) *There exist $c_1, \theta > 0$ such that for all $f \in \mathcal{F} \cap \mathbb{L}^1$,*

$$\|f\|_2^{2+4/\theta} \leq c_1(\mathcal{E}(f,f) + \delta\|f\|_2^2)\|f\|_1^{4/\theta}. \tag{3.2}$$

(ii) *For all $t > 0$, $P_t(\mathbb{L}^1) \subset \mathbb{L}^\infty$ and it is a bounded operator. Moreover, there exist $c_2, \theta > 0$ such that*

$$\|P_t\|_{1\to\infty} \leq c_2 e^{\delta t} t^{-\theta/2}, \qquad \forall t > 0. \tag{3.3}$$

Here $\|P_t\|_{1\to\infty}$ is the operator norm of $P_t : \mathbb{L}^1 \to \mathbb{L}^\infty$.

When $\delta = 0$, we cite (3.2) as (N_θ) and (3.3) as (UC_θ).

Equation (3.3) implies the existence of the heat kernel; there exists $\mathcal{N}$ with $\mu(\mathcal{N}) = 0$ and a kernel $p_t(x, y)$ of P_t (with respect to μ) defined on $(0, \infty) \times (X \setminus \mathcal{N}) \times (X \setminus \mathcal{N})$ such that

$$p_t(x, y) \leq c_1 t^{-\theta/2} e^{\delta t} \qquad \forall t > 0, \ \forall x, y \in X \setminus \mathcal{N}.$$

(In fact, one can choose $\mathcal{N}$ to be an "exceptional set"; see [22, Theorem 1.2].)

Proof. First, note that using Lemma 3.1.3, it is enough to prove the theorem when $\delta = 0$.

$(i) \Rightarrow (ii)$: Let $f \in \mathbb{L}^2 \cap \mathbb{L}^1$ with $\|f\|_1 = 1$ and $u(t) := (P_t f, P_t f)$. Then, by Lemma 3.1.1 (ii), $u'(t) = -2\mathcal{E}(P_t f, P_t f)$. Now by (i) and Lemma 3.1.1 (i),

$$2u(t)^{1+2/\theta} \le c_1(-u'(t))\|P_t f\|_1^{4/\theta} \le -c_1 u'(t),$$

so $u'(t) \le -c_2 u(t)^{1+2/\theta}$. Set $v(t) = u(t)^{-2/\theta}$, then we obtain $v'(t) \ge 2c_2/\theta$. Since $\lim_{t \downarrow 0} v(t) = u(0)^{-2/\theta} = \|f\|_2^{-4/\theta} > 0$, it follows that $v(t) \ge 2c_2 t/\theta$. This means $u(t) \le c_3 t^{-\theta/2}$, whence $\|P_t f\|_2 \le c_3 t^{-\theta/4} \|f\|_1$ for all $f \in \mathbb{L}^2 \cap \mathbb{L}^1$, which implies $\|P_t\|_{1 \to 2} \le c_3 t^{-\theta/4}$. Since $P_t = P_{t/2} \circ P_{t/2}$ and $\|P_{t/2}\|_{1 \to 2} = \|P_{t/2}\|_{2 \to \infty}$, we obtain (ii).

$(ii) \Rightarrow (i)$: Let $f \in \mathbb{L}^2 \cap \mathbb{L}^1$ with $\|f\|_1 = 1$. Using (ii) and Lemma 3.1.1 (iii), we have

$$\exp\left(-2\frac{\mathcal{E}(f,f)t}{\|f\|_2^2}\right) \le \frac{c_4 t^{-\theta/2}}{\|f\|_2^2}.$$

Rewriting, we have $\mathcal{E}(f,f)/\|f\|_2^2 \ge (2t)^{-1} \log(t^{\theta/2}\|f\|_2^2) - (2t)^{-1} A$, where $A = \log c_4$ and we may take $A > 0$. Set $\Psi(x) = \sup_{t>0}\{\frac{x}{2t}\log(xt^{\theta/2}) - Ax/(2t)\}$. By elementary computations, we have $\Psi(x) \ge c_5 x^{1+2/\theta}$. So

$$\mathcal{E}(f,f) \ge \Psi(\|f\|_2^2) \ge c_5 \|f\|_2^{2(1+2/\theta)} = c_5 \|f\|_2^{2+4/\theta}.$$

Since this holds for all $f \in \mathbb{L}^2 \cap \mathbb{L}^1$ with $\|f\|_1 = 1$, we obtain (i).   □

Remark 3.1.5. (1) If we are only concerned with $t \ge 1$ (for example on graphs), we have the following equivalence under the assumption of $\|P_t\|_{1 \to \infty} \le C$ for all $t \ge 0$ (C is independent of t).

   (i) (N_θ) with $\delta = 0$ holds for $\mathcal{E}(f,f) \le \|f\|_1^2$.
   (ii) (UC_θ) with $\delta = 0$ holds for $t \ge 1$.

(2) We have the following generalization of the theorem due to [76]. Let $m : \mathbb{R}_+ \to \mathbb{R}_+$ be a decreasing C^1 bijection which satisfies the following: $M(t) := -\log m(t)$ satisfies $M'(u) \ge c_0 M'(t)$ for all $t \ge 0$ and all $u \in [t, 2t]$. (Roughly, this condition means that the logarithmic derivative of $m(t)$ has polynomial growth. So exponential growth functions may satisfy the condition, but doubly exponential growth functions do not.) Let $\Psi(x) = -m'(m^{-1}(x))$. Then the following are equivalent.

   (i) $c_1 \Psi(\|f\|_2^2) \le \mathcal{E}(f,f)$ for all $f \in \mathcal{F}, \|f\|_1 \le 1$.
   (ii) $\|P_t\|_{1 \to \infty} \le c_2 m(t)$ for all $t > 0$.

The above theorem corresponds to the case $\Psi(y) = c_4 y^{1+2/\theta}$ and $m(t) = t^{-\theta/2}$.

Corollary 3.1.6. *Suppose the Nash inequality (Theorem 3.1.4) holds. Let φ be an eigenfunction of $-\mathcal{L}$ with eigenvalue $\lambda \ge 1$. Then*

$$\|\varphi\|_\infty \le c_1 \lambda^{\theta/4} \|\varphi\|_2,$$

where $c_1 > 0$ is a constant independent of φ and λ.

Proof. Since $-\mathcal{L}\varphi = \lambda\varphi$, $P_t\varphi = e^{t\mathcal{L}}\varphi = e^{-\lambda t}\varphi$. By Theorem 3.1.4, $\|P_t\|_{2\to\infty} = \|P_t\|_{1\to\infty}^{1/2} \le ct^{-\theta/4}$ for $t \le 1$. Thus

$$e^{-\lambda t}\|\varphi\|_\infty = \|P_t\varphi\|_\infty \le ct^{-\theta/4}\|\varphi\|_2.$$

Taking $t = \lambda^{-1}$ and $c_1 = ce$, we obtain the result. $\qquad\qquad\square$

Example 3.1.7. Consider $\mathbb{Z}^d$, $d \ge 2$ and put weight 1 for each edge $\{x, y\}$, $x, y \in \mathbb{Z}^d$ with $\|x - y\| = 1$. Then it is known that the corresponding simple random walk enjoys the following heat kernel estimate: $p_n(x, y) \le c_1 n^{-d/2}$ for all $x, y \in \mathbb{Z}^d$ and all $n \ge 1$. Now, let

$$H = \{\{(2n_1, 2n_2, \cdots, 2n_d), (2n_1 + 1, 2n_2, \cdots, 2n_d)\} : n_1, \cdots, n_d \in \mathbb{Z}\}$$

and consider a random subgraph $\mathcal{C}(\omega)$ by removing each $e \in H$ with probability $p \in [0, 1]$ independently. (So the set of vertices of $\mathcal{C}(\omega)$ is $\mathbb{Z}^d$. Here ω is the randomness of the environments.) If we define the quadratic forms for the original graph and $\mathcal{C}(\omega)$ by $\mathcal{E}(\cdot, \cdot)$ and $\mathcal{E}^\omega(\cdot, \cdot)$ respectively, then it is easy to see that $\mathcal{E}^\omega(f, f) \le \mathcal{E}(f, f) \le 4\mathcal{E}^\omega(f, f)$ for all $f \in \mathbb{L}^2$. Thus, by Remark 3.1.5 (1), the heat kernel of the simple random walk on $\mathcal{C}(\omega)$ still enjoys the estimate $p_n^\omega(x, y) \le c_1 n^{-d/2}$ for all $x, y \in \mathcal{C}(\omega)$ and all $n \ge 1$, for almost every ω.

3.2   The Faber-Krahn, Sobolev and Isoperimetric Inequalities

In this section, we denote $\Omega \subset\subset X$ when Ω is an open relative compact subset of X. (For weighted graphs, it simply means that Ω is a finite subset of X.) Let $C_0(X)$ be the space of continuous, compactly supported functions on X. Define

$$\lambda_1(\Omega) = \inf_{\substack{f \in \mathcal{F} \cap C_0(X), \\ \text{Supp } f \subset Cl(\Omega), f \ne \text{const}}} \frac{\mathcal{E}(f, f)}{\|f\|_2^2}, \qquad \forall \Omega \subset\subset X.$$

By the min-max principle, this is the first eigenvalue for the corresponding Laplace operator which is zero outside Ω.

Definition 3.2.1 (The Faber-Krahn Inequality). Let $\theta > 0$. We say $(\mathcal{E}, \mathcal{F})$ satisfies the Faber-Krahn inequality of order θ if the following holds:

$$\lambda_1(\Omega) \ge c\mu(\Omega)^{-2/\theta}, \qquad \forall \Omega \subset\subset X. \qquad\qquad (FK(\theta))$$

Theorem 3.2.2. $(N_\theta) \Leftrightarrow (FK(\theta))$

Proof. $(N_\theta) \Rightarrow (FK(\theta))$: This is an easy direction. From (N_θ), we have

$$\|f\|_2^2 \le c_1 \left(\frac{\|f\|_1^2}{\|f\|_2^2} \right)^{2/\theta} \mathcal{E}(f, f), \qquad \forall f \in \mathcal{F} \cap \mathbb{L}^1. \tag{3.4}$$

On the other hand, if Supp $f \subset Cl(\Omega)$, then by the Schwarz inequality, $\|f\|_1^2 \le \mu(\Omega)\|f\|_2^2$, so $(\|f\|_1^2/\|f\|_2^2)^{2/\theta} \le \mu(\Omega)^{2/\theta}$. Putting this into (3.4), we obtain $(FK(\theta))$.

$(FK(\theta)) \Rightarrow (N_\theta)$: We adopt the argument originated in [117]. Let $u \in \mathcal{F} \cap C_0(X)$ be a non-negative function. For each $\lambda > 0$, since $u < 2(u-\lambda)$ on $\{u > 2\lambda\}$, we have

$$\int u^2 d\mu = \int_{\{u>2\lambda\}} u^2 d\mu + \int_{\{u \le 2\lambda\}} u^2 d\mu$$

$$\le 4 \int_{\{u>2\lambda\}} (u-\lambda)^2 d\mu + 2\lambda \int_{\{u \le 2\lambda\}} u d\mu \le 4 \int (u-\lambda)_+^2 d\mu + 2\lambda \|u\|_1.$$

$$\tag{3.5}$$

Note that $(u - \lambda)_+ \in \mathcal{F}$ since $(\mathcal{E}, \mathcal{F})$ is Markovian (cf. [108, Theorem 1.4.1]). Set $\Omega = \{u > \lambda\}$; then Ω is an open relatively compact set since u is compactly supported, and Supp $(u - \lambda)_+ \subset \Omega$. So, applying $(FK(\theta))$ to $(u - \lambda)_+$ gives

$$\int (u-\lambda)_+^2 d\mu \le c_2 \mu(\Omega)^{2/\theta} \mathcal{E}((u-\lambda)_+, (u-\lambda)_+) \le c_2 \left(\frac{\|u\|_1}{\lambda} \right)^{2/\theta} \mathcal{E}(u, u),$$

where we used the Chebyshev inequality in the second inequality. Putting this into (3.5),

$$\|u\|_2^2 \le 4c_2 \left(\frac{\|u\|_1}{\lambda} \right)^{2/\theta} \mathcal{E}(u, u) + 2\lambda \|u\|_1.$$

Optimizing the right hand side by taking $\lambda = c_3 \mathcal{E}(u, u)^{\theta/(\theta+2)} \|u\|_1^{(2-\theta)/(2+\theta)}$, we obtain

$$\|u\|_2^2 \le c_4 \mathcal{E}(u, u)^{\frac{\theta}{\theta+2}} \|u\|_1^{\frac{4}{2+\theta}},$$

and thus obtain (N_θ). For general compactly supported $u \in \mathcal{F}$, we can obtain (N_θ) for u_+ and u_-, so for u as well. For general $u \in \mathcal{F} \cap \mathbb{L}^1$, approximation by compactly supported functions gives the desired result. $\qquad \square$

Remark 3.2.3. We can generalize Theorem 3.2.2 as follows. Let

$$\lambda_1(\Omega) \ge \frac{c}{\varphi(\mu(\Omega))^2}, \qquad \forall \Omega \subset\subset X, \tag{3.6}$$

where $\varphi : (0, \infty) \to (0, \infty)$ is a non-decreasing function. Then, it is equivalent to Remark 3.1.5(2)(i), where $\Psi(x) = x/\varphi(1/x)^2$, or $\varphi(x) = (x\Psi(1/x))^{-1/2}$. Theorem 3.2.2 is the case $\varphi(x) = x^{1/\theta}$.

In the following, we define $\|\nabla f\|_1$ for the two cases.

Case 1: When one can define the gradient on the space and $\mathcal{E}(f, f) = \frac{1}{2}\int_X |\nabla f(x)|^2 d\mu(x)$, then $\|\nabla f\|_1 := \int_X |\nabla f(x)|d\mu(x)$.

Case 2: When (X, μ) is a weighted graph, then $\|\nabla f\|_1 := \frac{1}{2}\sum_{x,y \in X} |f(y) - f(x)|\mu_{xy}$.

Whenever $\|\nabla f\|_1$ appears, we are one of the two cases.

Definition 3.2.4 (Sobolev Inequalities).

(i) Let $\theta > 2$. We say $(\mathcal{E}, \mathcal{F})$ satisfies (S_θ^2) if

$$\|f\|_{2\theta/(\theta-2)}^2 \le c_1 \mathcal{E}(f, f), \qquad \forall f \in \mathcal{F} \cap C_0(X). \qquad (S_\theta^2)$$

(ii) Let $\theta > 1$. We say $(\mathcal{E}, \mathcal{F})$ satisfies (S_θ^1) if

$$\|f\|_{\theta/(\theta-1)} \le c_2 \|\nabla f\|_1, \qquad \forall f \in \mathcal{F} \cap C_0(X). \qquad (S_\theta^1)$$

In the following, with some abuse of notation, we define $|\partial\Omega|$ for the two cases.

Case 1: When X is a d-dimensional Riemannian manifold and Ω is a smooth domain, then $|\partial\Omega|$ is the surface measure of Ω.

Case 2: When (X, μ) is a weighted graph, then $|\partial\Omega| = \sum_{x \in \Omega}\sum_{y \in X \setminus \Omega}\mu_{xy}$.

Whenever $|\partial\Omega|$ appears, we are one of the two cases.

Definition 3.2.5 (The Isoperimetric Inequality). Let $\theta > 1$. We say $(\mathcal{E}, \mathcal{F})$ satisfies the isoperimetric inequality of order θ if

$$\mu(\Omega)^{(\theta-1)/\theta} \le c_1|\partial\Omega|, \qquad \forall\Omega \subset\subset X. \qquad (I_\theta)$$

We write (I_∞) when $\theta = \infty$, namely when

$$\mu(\Omega) \le c_1|\partial\Omega|, \qquad \forall\Omega \subset\subset X. \qquad (I_\infty)$$

Remark 3.2.6. (i) For the weighted graph (X, μ) with $\mu_x \ge 1$ for all $x \in X$, if (I_β) holds, then (I_α) holds for any $\alpha \le \beta$. So (I_∞) is the strongest inequality among all the isoperimetric inequalities.

(ii) $\mathbb{Z}^d$ satisfies (I_d). The binary tree satisfies (I_∞).

Theorem 3.2.7. *The following holds for $\theta > 0$.*

$$(I_\theta) \overset{\theta>1}{\Longleftrightarrow} (S_\theta^1) \overset{\theta>2}{\Longrightarrow} (S_\theta^2) \overset{\theta>2}{\Longleftrightarrow} (N_\theta) \Longleftrightarrow (UC_\theta) \Longleftrightarrow (FK(\theta))$$

Proof. Note that the last two equivalence relations are already proved in Theorems 3.1.4 and 3.2.2.

$(I_\theta) \overset{\theta>1}{\Longleftarrow} (S^1_\theta)$: When (X, μ) is the weighted graph, simply apply (S^1_θ) to $f = 1_\Omega$ and we can obtain (I_θ). When X is the Riemannian manifold, by using a Lipschitz function which approximates $f = 1_\Omega$ nicely, we can obtain (I_θ).

$(I_\theta) \overset{\theta>1}{\Longrightarrow} (S^1_\theta)$: We will use the co-area formula given in Lemma 3.2.9 below. Let f be a support compact non-negative function on X (when X is the Riemannian manifold, $f \in C^1_0(X)$ and $f \geq 0$). Let $H_t(f) = \{x \in X : f(x) > t\}$ and set $p = \theta/(\theta - 1)$. Applying (I_θ) to f and using Lemma 3.2.9 below, we have

$$\|\nabla f\|_1 = \int_0^\infty |\partial H_t(f)| dt \geq c_1 \int_0^\infty \mu(H_t(f))^{1/p} dt = c_1 \int_0^\infty \|1_{H_t(f)}\|_p dt.$$
(3.7)

Next take any $g \in \mathbb{L}^q$ such that $g \geq 0$ and $\|g\|_q = 1$ where q is the value that satisfies $p^{-1} + q^{-1} = 1$. Then, by the Hölder inequality,

$$\int_0^\infty \|1_{H_t(f)}\|_p dt \geq \int_0^\infty \|g \cdot 1_{H_t(f)}\|_1 dt = \int_X g(x) \int_0^\infty 1_{H_t(f)}(x) dt d\mu(x) = \|fg\|_1,$$

since $\int_0^\infty 1_{H_t(f)}(x) dt = f(x)$. Putting this into (3.7), we obtain

$$\|f\|_p = \sup_{g \in \mathbb{L}^q : \|g\|_q = 1} \|fg\|_1 \leq c_1^{-1} \|\nabla f\|_1,$$

so we have (S^1_θ). We can obtain (S^1_θ) for general $f \in \mathcal{F} \cap C_0(X)$ by approximations.

$(S^1_\theta) \overset{\theta>2}{\Longrightarrow} (S^2_\theta)$: Set $\hat{\theta} = 2(\theta - 1)/(\theta - 2)$ and let $f \in \mathcal{F} \cap C_0(X)$. Applying (S^1_θ) to $f^{\hat{\theta}}$ and using the Schwarz inequality,

$$\left(\int f^{\frac{2\theta}{\theta-2}} d\mu \right)^{\frac{\theta-1}{\theta}} = \|f^{\hat{\theta}}\|_{\frac{\theta}{\theta-1}} \leq c_1 \|\nabla f^{\hat{\theta}}\|_1$$

$$\leq c_2 \|f^{\hat{\theta}-1} \nabla f\|_1 \leq c_2 \|\nabla f\|_2 \|f^{\hat{\theta}-1}\|_2$$

$$= c_2 \|\nabla f\|_2 \left(\int f^{\frac{2\theta}{\theta-2}} d\mu \right)^{1/2}.$$

rearranging, we obtain (S^2_θ).

$(S^2_\theta) \overset{\theta>2}{\Longrightarrow} (N_\theta)$: For $f \in \mathcal{F} \cap C_0(X)$, applying the Hölder inequality (with $p^{-1} = 4/(\theta + 2), q^{-1} = (\theta - 2)/(\theta + 2))$ and using (S^2_θ), we have

$$\|f\|_2^{2+\frac{4}{\theta}} \leq \|f\|_1^{\frac{4}{\theta}} \|f\|_{\frac{2\theta}{\theta-2}}^2 \leq c_1 \|f\|_1^{\frac{4}{\theta}} \mathcal{E}(f, f),$$

so we have (N_θ) in this case. Usual approximation arguments give the desired fact for $f \in \mathcal{F} \cap \mathbb{L}^1$.

$(S_\theta^2) \overset{\theta>2}{\Longleftarrow} (N_\theta)$: For $f \in \mathcal{F} \cap C_0(X)$ such that $f \geq 0$, define

$$f_k = (f - 2^k)_+ \wedge 2^k = 2^k 1_{A_k} + (f - 2^k) 1_{B_k}, \qquad k \in \mathbb{Z},$$

where $A_k = \{f \geq 2^{k+1}\}$, $B_k = \{2^k \leq f < 2^{k+1}\}$. Then $f = \sum_{k \in \mathbb{Z}} f_k$ and $f_k \in \mathcal{F} \cap C_0(X)$. So

$$\mathcal{E}(f, f) = \sum_{k \in \mathbb{Z}} \mathcal{E}(f_k, f_k) + \sum_{k \in \mathbb{Z}, k \neq k'} \sum_{k' \in \mathbb{Z}} \mathcal{E}(f_k, f_{k'}) \geq \sum_{k \in \mathbb{Z}} \mathcal{E}(f_k, f_k), \qquad (3.8)$$

where the last inequality is due to $\sum_{k \neq k'} \mathcal{E}(f_k, f_{k'}) \geq 0$. (This can be verified in a elementary way for the case of weighted graphs. When $\mathcal{E}$ is strongly local, $\sum_{k \neq k'} \mathcal{E}(f_k, f_{k'}) = 0$.)

Next, we have

$$\left(2^{2k} \mu(A_k)\right)^{1+2/\theta} = \left(\int_{A_k} f_k^2 d\mu\right)^{1+2/\theta}$$

$$\leq \|f_k\|_2^{2+4/\theta} \leq c_1 \|f_k\|_1^{4/\theta} \mathcal{E}(f_k, f_k)$$

$$\leq c_1 \left(2^k \mu(A_{k-1})\right)^{4/\theta} \mathcal{E}(f_k, f_k), \qquad (3.9)$$

where we used (N_θ) for f_k in the second inequality. Let $\alpha = 2\theta/(\theta - 2)$, $\beta = \theta/(\theta + 2) \in (1/2, 1)$, and define $a_k = 2^{\alpha k} \mu(A_k)$, $b_k = \mathcal{E}(f_k, f_k)$. Then (3.9) can be rewritten as $a_k \leq c_2 a_{k-1}^{2(1-\beta)} b_k^\beta$. Summing over $k \in \mathbb{Z}$ and using the Hölder inequality (with $p^{-1} = 1 - \beta$, $q^{-1} = \beta$), we have

$$\sum_k a_k \leq c_2 \sum_k a_{k-1}^{2(1-\beta)} b_k^\beta \leq c_2 (\sum_k a_{k-1}^2)^{1-\beta} (\sum_k b_k)^\beta$$

$$\leq c_2 (\sum_k a_k)^{2(1-\beta)} (\sum_k b_k)^\beta.$$

Putting (3.8) into this, we have

$$\sum_k a_k \leq c_2 \mathcal{E}(f, f)^{\beta/(2\beta - 1)}. \qquad (3.10)$$

On the other hand, we have

$$\|f\|_\alpha^\alpha = \sum_k \int_{B_k} f^\alpha d\mu \leq \sum_k 2^{\alpha(k+1)} \mu(A_{k-1}) = 2^{4\alpha} \sum_k a_k.$$

Plugging (3.10) into this, we obtain (S_θ^2). $\qquad\square$

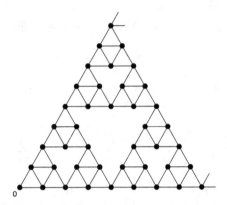

Fig. 3.1 Two-dimensional pre-Sierpinski gasket

Remark 3.2.8. (i) Generalizations of (S_θ^p) and (I_θ) are the following:

$$\|f\|_p \le c_1 \varphi(\mu(\Omega))\|\nabla f\|_p, \quad \forall \Omega \subset\subset X, \ \forall f \in \mathcal{F} \cap C_0(X), \ \text{Supp} \ f \subset Cl(\Omega),$$

$$\frac{1}{\varphi(\mu(\Omega))} \le c_2 \frac{|\partial \Omega|}{\mu(\Omega)}, \qquad \forall \Omega \subset\subset X,$$

where $\varphi : (0, \infty) \to (0, \infty)$ is a non-decreasing function as in Remark 3.2.3. Note that we defined (S_θ^p) for $p = 1, 2$, but the above generalization makes sense for all $p \in [1, \infty]$ (at least for Riemannian manifolds). Theorem 3.2.7 is the case $\varphi(x) = x^{1/\theta}$—see [75] for details.

(ii) As we see in the proof, in addition to the equivalence $(I_\theta) \iff (S_\theta^1)$, one can see that the best constant for c_1 in (I_θ) is equal to the best constant for c_2 in (S_θ^1). This is sometimes referred to as the Federer-Fleming theorem.

(iii) It is easy to see that the pre-Sierpinski gasket (Fig. 3.1) does not satisfy (I_θ) for any $\theta > 1$. On the other hand, we will prove in (4.29) that the heat kernel of the simple random walk enjoys the following estimate $p_{2n}(0,0) \asymp n^{-\log 3/\log 5}$, $\forall n \ge 1$. This gives an example that (N_θ) cannot imply (I_θ) in general. In other words, the best exponent for isoperimetric inequalities is not necessarily the best exponent for Nash inequalities.

(iv) In Sect. 3.4, we will introduce an interesting approach due to Morris and Peres [182] to prove $(I_\theta) \Rightarrow (N_\theta)$ directly using the evolution of random sets (see Proposition 3.4.3).

The following lemma was used in the proof of Theorem 3.2.7.

Lemma 3.2.9 (Co-area Formula). *Let f be a non-negative function on X (when X is the Riemannian manifold, $f \in C_0^1(X)$ and $f \ge 0$), and define $H_t(f) = \{x \in X : f(x) > t\}$. Then*

$$\|\nabla f\|_1 = \int_0^\infty |\partial H_t(f)|\, dt.$$

Proof. For simplicity we will prove it only when (X, μ) is a weighted graph. Then,

$$\|\nabla f\|_1 = \frac{1}{2} \sum_{x,y \in X} |f(y) - f(x)| \mu_{xy} = \sum_{x \in X} \sum_{y \in X: f(y) > f(x)} (f(y) - f(x)) \mu_{xy}$$

$$= \sum_{x \in X} \sum_{y \in X: f(y) > f(x)} \left(\int_0^\infty 1_{\{f(y) > t \geq f(x)\}} dt \right) \mu_{xy}$$

$$= \int_0^\infty dt \sum_{x,y \in X} 1_{\{f(y) > t \geq f(x)\}} \mu_{xy} = \int_0^\infty dt \sum_{y \in H_t(f)} \sum_{x \in H_t(f)^c} \mu_{xy}$$

$$= \int_0^\infty |\partial H_t(f)| dt.$$

$\square$

The next fact is an immediate corollary to Theorem 3.2.7.

Corollary 3.2.10. *Let $\theta > 2$. If $(\mathcal{E}, \mathcal{F})$ satisfies (I_θ), then the following holds.*

$$p_t(x, y) \leq ct^{-\theta/2} \qquad \forall t > 0, \ \mu\text{-a.e. } x, y.$$

3.3 Poincaré Inequalities and Relative Isoperimetric Inequalities

We first introduce some volume growth properties and Poincaré inequalities. Let (X, μ) be a weighted graph and for each $x \in X, R \geq 1$, set

$$V(x, R) := \mu(B(x, R))$$

where $B(x, R)$ is the ball of radius R centered at x with respect to the graph distance.

Definition 3.3.1. (1) We say (X, μ) satisfies *(VD)*, a volume doubling condition, if there exists $C_1 > 1$ such that

$$V(x, 2R) \leq C_1 V(x, R) \qquad \text{for all } x \in X, R \geq 1.$$

(2) We say (X, μ) satisfies *(V(d))*, a d-set condition, if there exists $C_2 \geq 1$ such that

$$C_2^{-1} R^d \leq V(x, R) \leq C_2 R^d \qquad \text{for all } x \in X, R \geq 1.$$

(3) We say (X, μ) satisfies $(PI(\beta))$, a scaled (strong) Poincaré inequality with parameter $\beta > 0$, if there exists a constants $C_3 > 0$ such that for any $B = B(x_0, R) \subset X$ with $x_0 \in X, R \geq 1$ and $f : B \to \mathbb{R}$,

$$\sum_{x \in B} (f(x) - \bar{f}_B)^2 \mu_x \leq C_3 R^\beta \sum_{x, y \in B} \mu_{xy} (f(x) - f(y))^2,$$

where $\bar{f}_B = \mu(B)^{-1} \sum_{y \in B} f(y) \mu_y$.

(4) We say (X, μ) satisfies $(WPI(\beta))$, a scaled weak Poincaré inequality with parameter $\beta > 0$, if there exists a constant $C_4 > 0$ such that for any $B(x_0, 2R) \subset X$ with $x_0 \in X, R \geq 1$ and $f : B(x_0, 2R) \to \mathbb{R}$,

$$\sum_{x \in B(x_0, R)} (f(x) - \bar{f}_{B(x_0, R)})^2 \mu_x \leq C_4 R^\beta \sum_{x, y \in B(x_0, 2R)} \mu_{xy} (f(x) - f(y))^2.$$

The Poincaré inequality $(PI(\beta))$ is sometimes called a *spectral gap inequality* since it implies that the spectral gap for $B(x_0, R)$ is at least $cR^{-\beta}$.

In the definition of $(WPI(\beta))$, the sum on the right hand side is over all $x, y \in B(x_0, 2R)$ whereas in $(PI(\beta))$, it is over all $x, y \in B(x_0, R)$. Clearly, $(PI(\beta))$ implies $(WPI(\beta))$. The next proposition asserts that if (VD) holds, the converse is true.

Proposition 3.3.2. *Let (X, μ) be a weighted graph that has controlled weights. Then*

$$(VD) + (WPI(\beta)) \Rightarrow (PI(\beta)).$$

This result was first proved by Jerison [142] under the setting of Euclidean vector fields satisfying Hörmander's condition—see [129] and [196] for more general formulation. For the graph setting, we refer the reader to the Appendix of [20] for the proof.

Under a suitable volume growth condition such as $(V(d))$, the Poincaré inequality implies the Nash inequality.

Proposition 3.3.3. *Let (X, μ) be a weighted graph that has controlled weights. Then*

$$(V(d)) + (WPI(\beta)) \Rightarrow (N_{2d/\beta}).$$

This proposition can be proved along the same lines as in the proof of [196, Sect. 3.3]. We will skip the proof.

However, the Nash inequality does not imply the Poincaré inequality even if $(V(d))$ is assumed. Indeed, let $\mathbb{Z}^3 \sharp \mathbb{Z}^3$ be a join of two $\mathbb{Z}^3$ constructed as follows; vertex set of $\mathbb{Z}^3 \sharp \mathbb{Z}^3$ is $\mathbb{Z}^3_{(1)} \cup \mathbb{Z}^3_{(2)}$ where $\mathbb{Z}^3_{(i)}, i = 1, 2$ are two copies of $\mathbb{Z}^3$, and the edge set of $\mathbb{Z}^3 \sharp \mathbb{Z}^3$ is $\mathbb{E}^3_{(1)} \cup \mathbb{E}^3_{(2)} \cup \{x_1, x_2\}$, where $\mathbb{E}^3_{(i)}, i = 1, 2$ are copies of the set of edges in $\mathbb{Z}^3$, and $x_i \in \mathbb{Z}^3_{(i)}$, $i = 1, 2$. We put conductance 1 on each edge of $\mathbb{Z}^3 \sharp \mathbb{Z}^3$. It can be easily checked that (N_3) and $(V(3))$ hold for this graph. But $(PI(2))$ does not hold, because if we let $f(x) = 1_{\mathbb{Z}^3_{(1)}}(x) - 1_{\mathbb{Z}^3_{(2)}}(x)$, then setting $B = B(x_1, R)$, we have

$$\sum_{x \in B} (f(x) - \bar{f}_B)^2 \asymp R^3, \quad \frac{1}{2} \sum_{x,y \in B} (f(x) - f(y))^2 = 4.$$

We note that in [119], detailed heat kernel estimates for Brownian motion on a join of $\mathbb{R}^m$ and $\mathbb{R}^n$ are given.

Scaled Poincaré inequalities are very useful since they imply detailed heat kernel estimates under a mild volume growth condition as we will see in Theorem 3.3.5.

Definition 3.3.4. (1) We say (X, μ) satisfies the sub-Gaussian (Gaussian, when $\beta = 2$) heat kernel estimates of order β if whenever $x, y \in X, n \geq d(x, y) \vee 1$, the heat kernel of $\{Y_n\}_n$ enjoys the following estimates:

$$p_n(x, y) \leq \frac{C_1}{V(x, n^{1/\beta})} \exp[-(\frac{d(x, y)^\beta}{C_1 n})^{1/(\beta-1)}], \qquad (UHKE(\beta))$$

$$p_n(x, y) + p_{n+1}(x, y) \geq \frac{C_2}{V(x, n^{1/\beta})} \exp[-(\frac{d(x, y)^\beta}{C_2 n})^{1/(\beta-1)}]. \quad (LHKE(\beta))$$

(2) We say (X, μ) satisfies $(PHI(\beta))$, a parabolic Harnack inequality of order β if whenever $u(n, x) \geq 0$ is defined on $[0, 4N] \times B(z, 2R+1)$ where $z \in X, N \geq 1$, $R \geq 1$, and satisfies

$$u(n + 1, x) - u(n, x) = \mathcal{L}u(n, x), \qquad (n, x) \in [0, 4N] \times B(z, 2R),$$

then

$$\max_{\substack{N \leq n \leq 2N \\ x \in B(z,R)}} u(n, x) \leq C_1 \min_{\substack{3N \leq n \leq 4N \\ x \in B(z,R)}} (u(n, x) + u(n + 1, x)), \qquad (3.11)$$

where $N \geq 2R$ and $N \asymp R^\beta$ (Fig. 3.2).

In the definition, a priori $\beta > 1$ but in fact, the estimates $(UHKE(\beta))$ can hold only if $\beta \geq 2$. One way to see this is to observe that the upper bound $p_n(x, x) \leq \frac{C}{V(x,n^{1/\beta})}$ which follows from $(UHKE(\beta))$ is compatible with the lower bound from [173], $p_n(x, x) \geq \frac{c}{V(x,n^{1/2}\log n)}$, which always holds under (VD), only if $\beta \geq 2$. By integrating $(UHKE(\beta)) + (LHKE(\beta))$, $\{Y_n\}_n$ has the following sub-diffusive behavior, and that is why the heat kernel estimate is called "sub-Gaussian" when $\beta > 2$:

$$\mathbb{E}^x[d(x, Y_n)] \asymp n^{1/\beta} \leq n^{1/2}.$$

As in Remark 4.5.3 below, $(UHKE(\beta)) + (LHKE(\beta))$ holds with some $\beta > 2$ for simple random walks on various fractal graphs.

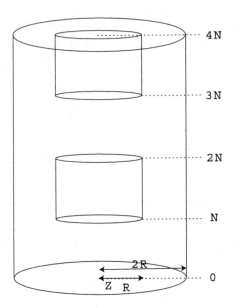

Fig. 3.2 Parabolic cylinder

When $\beta = 2$, the behavior is diffusive, and under $(V(d))$ the heat kernel is bounded from above and below by Gaussian functions $\frac{c_1}{n^{d/2}} \exp(-\frac{c_2 d(x,y)^2}{n})$ up to constants, and that is why the heat kernel estimate is called "Gaussian".

Under $(V(d))$, the estimates $(UHKE(\beta)) + (LHKE(\beta))$ can hold only if $\beta \leq d + 1$. This can be seen in several ways, for example because the lower bound $p_n(x, x) \geq c\,n^{-d/\beta}$ which follows from $(LHKE(\beta))$ must be compatible with the upper bound $p_n(x, x) \leq C\,n^{-d/(d+1)}$, which will be shown in Proposition 4.3.2. Conversely, it was proved in [17] that for every couple d, β such that $2 \leq \beta \leq d+1$, there exists a graph (X, μ) with $(V(d))$ such that $(UHKE(\beta)) + (LHKE(\beta))$ holds.

The following important equivalence for $\beta = 2$ was proved in [89], based on the corresponding results for the case of manifolds in [116, 194] and for the case of metric measure spaces in [207]. (To be precise, the second equivalence in (3.12) was known before, at least implicitly. See for example, [99].)

Theorem 3.3.5. *Let (X, μ) be a weighted graph that has controlled weights. Then*

$$(VD) + (PI(2)) \Leftrightarrow (UHKE(2)) + (LHKE(2)) \Leftrightarrow (PHI(2)). \qquad (3.12)$$

It can be checked that both (VD) and $(PI(2))$ are stable under rough isometry (thus, stable under bounded perturbation). So, one can see that the Gaussian heat kernel estimates and the parabolic Harnack inequality of order 2 are both stable under rough isometry.

When $\beta \neq 2$, the first equivalence of (3.12) no longer holds $((VD)+(PI(2))$ does not imply $(UHKE(2))+(LHKE(2)))$. In order to explain an alternative equivalence, we give one more definition.

Definition 3.3.6. Let $\beta \geq 2$. We say (X, μ) satisfies $(CS(\beta))$, a cut-off Sobolev inequality with exponent β, if there exist constants $\theta \in (0, 1]$ and $C_1, C_2 > 0$ such that for every $x_0 \in X$, $R \geq 1$, there exists a cut-off function $\varphi(= \varphi_{x_0,R})$ satisfying the following properties.

(a) $\varphi(x) \geq 1$ for $x \in B(x_0, R/2)$ and $\varphi(x) = 0$ for $x \in B(x_0, R)^c$.
(b) $|\varphi(x) - \varphi(y)| \leq C_1 (d(x, y)/R)^\theta$ for all $x, y \in X$.
(c) For any ball $B(x_0, s)$ with $1 \leq s \leq R$ and $f : B(x_0, 2s) \to \mathbb{R}$,

$$\sum_{x \in B(x_0,s)} f(x)^2 \sum_{y \in X} \mu_{xy} |\varphi(x) - \varphi(y)|^2 \tag{3.13}$$

$$\leq C_2 (s/R)^{2\theta} \Big(\sum_{x,y \in B(x_0,2s)} \mu_{xy} |f(x) - f(y)|^2 + s^{-\beta} \sum_{y \in B(x_0,2s)} f(y)^2 \mu_y \Big).$$

The following equivalence is proved in [21] (see [23, 30] for the case of metric measure spaces).

Theorem 3.3.7. *Let (X, μ) be a weighted graph that has controlled weights. Then*

$$(VD) + (PI(\beta)) + (CS(\beta)) \Leftrightarrow (UHKE(\beta)) + (LHKE(\beta)) \Leftrightarrow (PHI(\beta)). \tag{3.14}$$

Again, it can be checked that $(VD), (PI(\beta))$ and $(CS(\beta))$ are all stable under rough isometry (see [130]). So, one can see that the sub-Gaussian heat kernel estimates and the parabolic Harnack inequality of order β are both stable under rough isometry.

In a recent paper [9], the authors introduce a simplification of the condition $(CS(\beta))$, which they denote by $(CSA(\beta))$, where the Hölder continuity of the cut-off function φ is no longer assumed. They prove that under (VD), a version of the Faber-Krahn inequality plus $(CSA(\beta))$ is equivalent to $(UHKE(\beta))$ in the setting of metric measure spaces. It implies that $(UHKE(\beta))$ is stable under rough isometry. In the paper, they also point out that one can replace $(CS(\beta))$ in (3.14) to a simpler condition $(CSA(\beta))$.

Finally, following [20] we give another type of isoperimetric inequality that implies the Poincaré inequalities. For $A, B \subset X$, let

$$\mu(A; B) := \sum_{x \in A} \sum_{y \in B} \mu_{xy}.$$

Definition 3.3.8. X satisfies a relative isoperimetric inequality (RI) with constant C_R if the following holds; For any $x_0 \in X$, $R \geq 1$ and $A \subset B := B(x_0, R)$ with $\mu(A) \leq \frac{1}{2}\mu(B)$,

$$\frac{\mu(A; B \setminus A)}{\mu(A)} \geq C_R. \tag{3.15}$$

Note the difference between the two isoperimetric inequalities, namely, the relative isoperimetric inequality does not count the part which is the boundaries for both A and B. Thus

$$|\partial A| = \mu(A; X \setminus A) \geq \mu(A; B \setminus A). \tag{3.16}$$

The Cheeger constant for a finite graph is defined by

$$\Phi_X := \min\{\frac{\mu(A; X \setminus A)}{\mu(A)} : A \subset X, 0 < \mu(A) \leq \frac{1}{2}\mu(X)\}.$$

For a finite graph, if we let R_I be the largest constant that satisfies (3.15) for all $R \geq 1$, then noting (3.16), one can see $R_I = \Phi_X$.

For $f : B \to \mathbb{R}$, let $M_B(f)$ be a median of f, i.e. it satisfies

$$\mu(\{x \in B : f(x) > M_B(f)\}) \vee \mu(\{x \in B : f(x) < M_B(f)\}) \leq \frac{1}{2}\mu(B). \tag{3.17}$$

It is easy to see $\min_a \sum_{y \in B} |f(y) - a|\mu_y$ is attained by the median.

Proposition 3.3.9. *The following are equivalent.*

(a) *X satisfies (RI) with constant C_R for any $R \geq 1$.*
(b) *The following (1,1)-Poincaré inequality holds for any $f : B = B(x_0, R) \to \mathbb{R}$, any $x_0 \in X$ and $R \geq 1$,*

$$\sum_{y \in B} |f(y) - M_B(f)|\mu_y \leq C_R^{-1}\|\nabla f\|_{1,B}, \tag{3.18}$$

where $\|\nabla f\|_{1,B} := \frac{1}{2}\sum_{x,y \in B} |f(y) - f(x)|\mu_{xy}$.

Proof. (b) $\Rightarrow$ (a) is easy; simply take $f = 1_A$ in (b) for $A \subset B$ with $\mu(A) \leq \frac{1}{2}\mu(B)$, then $M_B(f) = 0$ and one can obtain the relative isoperimetric inequality. For (a) $\Rightarrow$ (b), first, we may assume $M_B(g) = 0$ by taking $g = f(y) - M_B(f)$. Second, by splitting $g = g_+ - g_-$ and computing separately, we may assume $g \geq 0$. Now, under (a) we have

$$\sum_{x \in B} g(x)\mu_x = \int_0^\infty \mu(\{x \in B : g(x) > t\})dt$$

$$\leq C_R^{-1}\int_0^\infty \mu(\{x \in B : g(x) > t\}; \{x \in B : g(x) \leq t\})dt$$

$$= C_R^{-1}\|\nabla g\|_{1,B}.$$

Here the relative isoperimetric inequality is used in the first inequality and Lemma 3.2.9 is used in the last equality. This implies (3.18).   □

Proposition 3.3.10. *(RI) with C_R implies the Poincaré inequality with exponent C_R^{-2}, namely*

$$\sum_{x \in B}(f(x) - \bar{f}_B)^2 \mu_x \leq C_R^{-2} \sum_{x,y \in B} \mu_{xy}(f(x) - f(y))^2, \quad \forall f : B \to \mathbb{R}. \quad (3.19)$$

Proof. By subtracting a constant, we may assume $M_B(f) = 0$. Let $g(x) = f(x)^2 \mathrm{sgn}(f(x))$; then $M_B(g) = 0$. By Proposition 3.3.9, we have

$$\|f\|_{2,B}^2 = \sum_{x \in B} |g(x)| \mu_x \leq \frac{1}{2C_R} \sum_{x,y \in B} |g(y) - g(x)| \mu_{xy}. \quad (3.20)$$

Noting that

$$|a^2 \mathrm{sgn}(a) - b^2 \mathrm{sgn}(b)| \leq |a - b|(|a| + |b|) \qquad \forall a, b \in \mathbb{R},$$

we have

$$\text{(RHS of (3.20))} \leq \frac{1}{2C_R} \sum_{x,y \in B} |f(y) - f(x)|(|f(x)| + |f(y)|) \mu_{xy}$$

$$= C_R^{-1} \sum_{x,y \in B} |f(y) - f(x)||f(x)| \mu_{xy}$$

$$\leq C_R^{-1} \Big(\sum_{x,y \in B} \mu_{xy}(f(x) - f(y))^2 \Big)^{1/2} \|f\|_{2,B}.$$

Dividing by $\|f\|_{2,B}$ and taking square, we obtain (3.19). $\qquad \Box$

3.4 Evolving Random Sets and Heat Kernel Bounds

In this section, we introduce the notion of evolving random sets due to Morris and Peres [182], following [19]. Let (X, μ) be a weighted graph. Let 2^X be the set of finite subsets of X. We define a 2^X-valued Markov chain $\{S_n : n \geq 0\}$ as follows. Let $\{U_k : k \geq 1\}$ be i.i.d. uniform distributed random variables on $[0, 1]$. Given $S_0, \cdots, S_n$, define

$$S_{n+1} = \{y \in X : \mu(y; S_n) \geq \mu_y U_{n+1}\}.$$

Let us observe how this process evolves.

(a) y is added to S_n only if $y \in \partial S_n$.
(b) If $y \in \mathrm{Int}(S_n)$ (which means whenever $x \sim y$, $x \in S_n$), then $y \in S_{n+1}$.
(c) If $S_n = \emptyset$, then $S_{n+1} = \emptyset$.

As we see, S_{n+1} differs from S_n only at the boundary points.

For $A \subset X$, let P_A be the law of $\{S_n\}$ started at $S_0 = A$. We then have the following.

Lemma 3.4.1.   (i) $P_{\{x\}}(y \in S_n) = p_n(x, y)\mu_x$.
(ii) $\{\mu(S_n) : n \geq 0\}$ is a martingale.
(iii) $p_{2n}(x, x) \leq (E_{\{x\}}\mu(S_n)^{1/2})^2/\mu_x^2$

Proof. (i) Since $P_{\{x\}}(y \in S_0) = \delta_{xy}$ and $\mu_x p_0(x, y) = \mu_x \delta_{xy}/\mu_y = \delta_{xy}$, the result holds when $n = 0$. Now suppose the result holds for n and for all $x, y \in X$. Then

$$P_{\{x\}}(y \in S_{n+1}) = E_{\{x\}}[P(U_{n+1} \leq \mu(y; S_n)/\mu_y)] = E_{\{x\}}[\mu(y; S_n)/\mu_y]$$

$$= E_{\{x\}}[\sum_z 1_{\{z \in S_n\}}\mu_{yz}/\mu_y] = \sum_z p_n(x, z)\mu_x \mu_{yz}/\mu_y$$

$$= \sum_z p_1(y, z)\mu_z p_n(z, x)\mu_x = p_{n+1}(x, y)\mu_x.$$

(ii) We have

$$E_{\{x\}}[\mu(S_{n+1})|S_n] = \sum_y \mu_y P_{\{x\}}(y \in S_{n+1}|S_n) = \sum_y \mu_y \frac{\mu(y; S_n)}{\mu_y}$$

$$= \sum_y \mu(y; S_n) = \mu(S_n).$$

(iii) Let $\{S_n : n \geq 0\}, \{S_n' : n \geq 0\}$ be independent copies of the process. We then have

$$p_{2n}(x, x) = \sum_y p_n(x, y)p_n(y, x)\mu_y = \sum_y P_{\{x\}}(y \in S_n)^2 \mu_y/\mu_x^2$$

$$= \sum_y P_{\{x\}}(y \in S_n)P_{\{x\}}(y \in S_n')\mu_y/\mu_x^2$$

$$= E_{\{x\}}[\sum_y 1_{\{y \in S_n\}}1_{\{y \in S_n'\}}\mu_y/\mu_x^2]$$

$$\leq E_{\{x\}}[(\sum_y 1_{\{y \in S_n\}}\mu_y)^{1/2}(\sum_y 1_{\{y \in S_n'\}}\mu_y)^{1/2}]/\mu_x^2$$

$$= E_{\{x\}}[\mu(S_n)^{1/2}\mu(S_n')^{1/2}]/\mu_x^2 = E_{\{x\}}[\mu(S_n)^{1/2}]^2/\mu_x^2.$$

So far we did not assume anything for the weighted graph.

Definition 3.4.2. A Markov chain $\{X_n : n \geq 1\}$ is called lazy if $P^x(X_1 = x) \geq 1/2$ for all x.

From now on, we assume that the Markov chain on (X, μ) is lazy and satisfies the isoperimetric inequality (I_θ). We will use Lemma 3.4.1 (iii) to prove (N_θ). (This was already proved in Theorem 3.2.7 without assuming the laziness of the Markov chain, but the proof below gives an alternative and direct proof of $(I_\theta) \Rightarrow (N_\theta)$.)

For non-void subset $A \subset X$, define

$$\gamma(A) = \mu(A; A^c)/\mu(A).$$

Then the isoperimetric inequality (I_θ) gives $\gamma(A) \geq c_0 \mu(A)^{-1/\theta}$.

Proposition 3.4.3. *Suppose the Markov chain on (X, μ) is lazy. Then (I_θ) implies (N_θ).*

Proof. Step 1: We claim that, for any non-void $A \subset X$,

$$E_A[\mu(S_1)^{1/2}] \leq \mu(A)^{1/2}(1 - c_1 \mu(A)^{-2/\theta}). \tag{3.21}$$

To prove this, let $S_0 = A$ and suppose $U_1 < 1/2$. If $y \in A$, then $\mu(y; A) \geq \mu_{yy} \geq \mu_y/2 > U_1 \mu_y$ (the lazy condition is used in the second inequality) so that $y \in S_1$. If $y \in \partial A$, then $y \in S_1$ if and only if $U_1 < \mu(y; A)/\mu_y$. So

$$P_A(y \in S_1 | U_1 < 1/2) = \begin{cases} 1 & \text{if } y \in A, \\ 2\mu(y; A)/\mu_y & \text{if } y \in \partial A. \end{cases}$$

Hence

$$E_A[\mu(S_1)|U_1 < 1/2] = \sum_y \mu_y P_A(y \in S_1 | U_1 < 1/2)$$

$$= \mu(A) + 2\mu(A; A^c) = \mu(A)(1 + 2\gamma(A)). \tag{3.22}$$

Since $\mu(S_n)$ is a martingale, $E_A[\mu(S_1)] = \mu(A)$. This and (3.22) implies

$$E_A[\mu(S_1)|U_1 > 1/2] = \mu(A) - 2\mu(A; A^c) = \mu(A)(1 - 2\gamma(A)).$$

Thus, using Jensen's inequality, we have

$$E_A[\mu(S_1)^{1/2}] = \frac{1}{2}E_A[\mu(S_1)^{1/2}|U_1 < 1/2] + \frac{1}{2}E_A[\mu(S_1)^{1/2}|U_1 > 1/2]$$

$$\leq \frac{1}{2}E_A[\mu(S_1)|U_1 < 1/2]^{1/2} + \frac{1}{2}E_A[\mu(S_1)|U_1 > 1/2]^{1/2}$$

$$= \frac{1}{2}(\mu(A)(1 + 2\gamma(A)))^{1/2} + \frac{1}{2}(\mu(A)(1 - 2\gamma(A)))^{1/2}$$

$$\leq \mu(A)^{1/2}(1 - \gamma(A)^2/2),$$

where we used the following inequality in the last part.

$$\frac{1}{2}(1+2t)^{1/2} + \frac{1}{2}(1-2t)^{1/2} \le (1-t^2)^{1/2} \le 1 - t^2/2 \qquad \text{for } |t| \le 1/2.$$

Using (I_θ), we obtain (3.21).

Step 2: We prove the following general inequality. Let X be a non-negative random variable with $EX = 1$ and $\delta \ge 0$. Then

$$E[X^{(1-\delta)/2}1_{\{X>0\}}] \ge (EX^{1/2})^{1+\delta}. \tag{3.23}$$

Indeed, let η, ξ be non-negative random variables with $E\eta = 1$, and define a new probability $\hat{P}$ by $\hat{E}[\xi] = E[\eta\xi]$. Applying Jensen's inequality,

$$E[\eta\xi^{1+p}] = \hat{E}\xi^{1+p} \ge (\hat{E}\xi)^{1+p} = E[\eta\xi]^{1+p}.$$

So, taking $\eta = X, \xi = X^{-1/2}1_{\{X>0\}}$ and $p = \delta$, we have

$$E[X^{(1-\delta)/2}1_{\{X>0\}}] = E[XX^{-(1+\delta)/2}1_{\{X>0\}}]$$
$$\ge (E[XX^{-1/2}1_{\{X>0\}}])^{1+\delta} = (EX^{1/2})^{1+\delta}.$$

Step 3: Now we prove the desired result. Let $Y_n = \mu(S_n)^{1/2}$ and $y_n = E_{\{x_0\}}[Y_n]/\mu_{x_0}$. Equation (3.21) gives

$$E_{\{x_0\}}[Y_{n+1}|Y_n] \le Y_n - c_1 Y_n^{1-4/\theta}1_{\{Y_n>0\}}.$$

Using this and (3.23) with $X = \mu(S_n) = Y_n^2$ (Note that by Lemma 3.4.1 (ii), $E_{\{x_0\}}\mu(S_n)/\mu_{x_0} = E_{\{x_0\}}\mu(S_0)/\mu_{x_0} = \mu_{x_0}/\mu_{x_0} = 1$), we have

$$y_{n+1} - y_n \le -c_1 E_{\{x_0\}}[Y_n^{1-4/\theta}1_{\{Y_n>0\}}]/\mu_{x_0}$$
$$= -c_1 E_{\{x_0\}}[X^{(1-4/\theta)/2}1_{\{X>0\}}]/\mu_{x_0} \le -c_1 y_n^{1+4/\theta}. \tag{3.24}$$

Note that (apart from the power) this is the discrete analogue of the differential inequality $u'(t) \le -c_2 u(t)^{1+2/\theta}$ in the proof of Theorem 3.1.4. Let $\delta = 4/\theta$. We have

$$y_{n+1} \le y_n(1 - c_1 y_n^\delta) \le y_n \exp(-c_1 y_n^\delta).$$

So, noting that y_n is decreasing (due to (3.24)),

$$\int_{y_{n+1}}^{y_n} t^{-1-\delta}dt \ge y_n^{-\delta}\log(y_n/y_{n+1}) \ge c_1.$$

Summing this from $n = 0$ to $n - 1$, we have

$$c_1 n \leq \int_{y_n}^{y_0} t^{-1-\delta} dt = \delta^{-1}(y_n^{-\delta} - y_0^{-\delta}) \leq \delta^{-1} y_n^{-\delta},$$

which implies $y_n \leq c_2 n^{-\theta/4}$ for $n \geq 1$. Rewriting, we have

$$y_n^2 = (E_{\{x_0\}}\mu(S_n)^{1/2})^2/\mu_{x_0}^2 \leq c_2^2 n^{-\theta/2}.$$

Plugging this into Lemma 3.4.1 (iii), we obtain the desired result. $\qquad\square$

Chapter 4
Heat Kernel Estimates Using Effective Resistance

In this chapter, we will study detailed asymptotic properties of Green functions and heat kernels using the effective resistance. Let (X, μ) be a weighted graph. We say that (X, μ) is a *tree* if for any $l \geq 3$, there is no set of distinct points $\{x_i\}_{i=1}^{l} \subset X$ such that $x_i \sim x_{i+1}$ for $1 \leq i \leq l$ where we set $x_{l+1} := x_1$.

Set $R_{\mathrm{eff}}(x, x) = 0$ for all $x \in X$. We now give an important lemma on the effective resistance.

Lemma 4.0.1. (i) If $c_1 := \inf_{x,y\in X:x\sim y} \mu_{xy} > 0$ then $R_{\mathrm{eff}}(x, y) \leq c_1^{-1}d(x, y)$ for all $x, y \in X$.

(ii) If (X, μ) is a tree and $c_2 := \sup_{x,y\in X:x\sim y} \mu_{xy} < \infty$, then $R_{\mathrm{eff}}(x, y) \geq c_2^{-1}d(x, y)$ for all $x, y \in X$.

(iii) $|f(x) - f(y)|^2 \leq R_{\mathrm{eff}}(x, y)\mathcal{E}(f, f)$ for all $x, y \in X$ and $f \in H^2$.

(iv) $R_{\mathrm{eff}}(\cdot, \cdot)$ and $R_{\mathrm{eff}}(\cdot, \cdot)^{1/2}$ are both metrics on X.

Proof. (i) Take a shortest path between x and y and cut all the bonds that are not along the path. Then we have the inequality by the cutting law.

(ii) Suppose $d(x, y) = n$. Take the shortest path $(x_0, x_1, \cdots, x_n)$ between x and y so that $x_0 = x, x_n = y$. Now take $f : X \to \mathbb{R}$ so that $f(x_i) = (n - i)/n$ for $0 \leq i < n$, and $f(z) = f(x_i)$ if z is in the branch from x_i, i.e. if y can be connected to x_i without crossing $\{x_k\}_{k=0}^{n} \setminus \{x_i\}$. This f is well-defined because (X, μ) is a tree, and $f(x) = 1, f(y) = 0$. So $R_{\mathrm{eff}}(x, y)^{-1} \leq (1/2) \cdot 2\sum_{i=0}^{n-1}(1/n)^2\mu_{x_i x_{i+1}} \leq c_2/n = c_2/d(x, y)$, and the result follows.

(iii) For any $u \in H^2$ and any $x \neq y \in X$ with $u(x) \neq u(y)$, we can construct $f \in H^2$ such that $f(x) = 1, f(y) = 0$ by a transform $f(z) = au(z) + b$ (where a, b are chosen suitably). So

$$\sup \left\{ \frac{|u(x) - u(y)|^2}{\mathcal{E}(u, u)} : u \in H^2, u(x) \neq u(y) \right\}$$

$$= \sup \left\{ \frac{1}{\mathcal{E}(f, f)} : f \in H^2, f(x) = 1, f(y) = 0 \right\} = R_{\mathrm{eff}}(x, y), \quad (4.1)$$

and we have the desired inequality.

T. Kumagai, *Random Walks on Disordered Media and their Scaling Limits*, Lecture Notes in Mathematics 2101, DOI 10.1007/978-3-319-03152-1_4,
© Springer International Publishing Switzerland 2014

(iv) It is easy to see $R_{\text{eff}}(x, y) = R_{\text{eff}}(y, x)$ and $R_{\text{eff}}(x, y) = 0$ if and only if $x = y$. So we only need to check the triangle inequality.

Let $\tilde{H}^2 = \{u \in H^2 : \mathcal{E}(u, u) > 0\}$. Then, for $x, y, z \in X$ that are distinct, we have by (4.1)

$$R_{\text{eff}}(x, y)^{1/2} = \sup\left\{\frac{|u(x) - u(y)|}{\mathcal{E}(u, u)^{1/2}} : u \in \tilde{H}^2\right\}$$

$$\leq \sup\left\{\frac{|u(x) - u(z)|}{\mathcal{E}(u, u)^{1/2}} : u \in \tilde{H}^2\right\} + \sup\left\{\frac{|u(z) - u(y)|}{\mathcal{E}(u, u)^{1/2}} : u \in \tilde{H}^2\right\}$$

$$= R_{\text{eff}}(x, z)^{1/2} + R_{\text{eff}}(z, y)^{1/2}.$$

So $R_{\text{eff}}(\cdot, \cdot)^{1/2}$ is a metric on X.

Next, let $V = \{x, y, z\} \subset X$ and let $\{\hat{\mu}_{xy}, \hat{\mu}_{yz}, \hat{\mu}_{zx}\}$ be the trace of $\{\mu_{xy}\}_{x,y \in X}$ to V. Define $R_{xy}^{-1} = \hat{\mu}_{xy}$, $R_{yz}^{-1} = \hat{\mu}_{yz}$, $R_{zx}^{-1} = \hat{\mu}_{zx}$. Then, using Proposition 2.3.1 and the resistance formula of series and parallel circuits, we have

$$R_{\text{eff}}(z, x) = \frac{1}{R_{zx}^{-1} + (R_{xy} + R_{yz})^{-1}} = \frac{R_{zx}(R_{xy} + R_{yz})}{R_{xy} + R_{yz} + R_{zx}}, \tag{4.2}$$

and similarly $R_{\text{eff}}(x, y) = \frac{R_{xy}(R_{yz}+R_{zx})}{R_{xy}+R_{yz}+R_{zx}}$ and $R_{\text{eff}}(y, z) = \frac{R_{yz}(R_{zx}+R_{xy})}{R_{xy}+R_{yz}+R_{zx}}$. Hence

$$\frac{1}{2}\{R_{\text{eff}}(x, z) + R_{\text{eff}}(z, y) - R_{\text{eff}}(x, y)\} = \frac{R_{yz}R_{zx}}{R_{xy} + R_{yz} + R_{zx}} \geq 0, \tag{4.3}$$

which shows that $R_{\text{eff}}(\cdot, \cdot)$ is a metric on X.   □

Remark 4.0.2. (i) Different proofs of the triangle inequality of $R_{\text{eff}}(\cdot, \cdot)$ in Lemma 4.0.1 (iv) can be found in [20] and [147, Theorem 1.12].

(ii) Weighted graphs are resistance forms. (See [151] for definition and properties of the resistance form.) In fact, most of the results in this chapter (including this lemma) hold for resistance forms.

4.1   Green Density on a Finite Set

For $y \in X$ and $n \in \mathbb{N}$, let

$$L(y, n) = \sum_{k=0}^{n-1} 1_{\{Y_k = y\}}$$

be the local time at y up to time $n - 1$. For a finite set $B \subset X$ and $x, y \in X$, define the Green density by

$$g_B(x, y) = \frac{1}{\mu_y} \mathbb{E}^x[L(y, \tau_B)] = \frac{1}{\mu_y} \sum_k \mathbb{P}^x(Y_k = y, k < \tau_B). \qquad (4.4)$$

Clearly $g_B(x, y) = 0$ when either x or y is outside B. Since $\mu_y^{-1} \mathbb{P}^x(Y_k = y, k < \tau_B) = \mu_x^{-1} \mathbb{P}^y(Y_k = x, k < \tau_B)$, we have

$$g_B(x, y) = g_B(y, x) \qquad \forall x, y \in X. \qquad (4.5)$$

Using the strong Markov property of Y,

$$g_B(x, y) = \mathbb{P}^x(\sigma_y < \tau_B) g_B(y, y) \le g_B(y, y). \qquad (4.6)$$

Below are further properties of the Green density.

Lemma 4.1.1. *Let $B \subset X$ be a finite set. Then the following hold.*

(i) *For $x \in B$, $g_B(x, \cdot)$ is harmonic on $B \setminus \{x\}$ and $= 0$ outside B.*
(ii) *(Reproducing property of the Green density) For all $f \in H^2$ with Supp $f \subset B$, it holds that $\mathcal{E}(g_B(x, \cdot), f) = f(x)$ for each $x \in X$.*
(iii) *$\mathbb{E}^x[\tau_B] = \sum_{y \in B} g_B(x, y) \mu_y$ for each $x \in X$.*
(iv) *$R_{\text{eff}}(x, B^c) = g_B(x, x)$ for each $x \in X$.*
(v) *$\mathbb{E}^x[\tau_B] \le R_{\text{eff}}(x, B^c) \mu(B)$ for each $x \in X$.*

Proof.   (i) Let $v(z) = g_B(x, z)$. Then, for each $y \in B \setminus \{x\}$, noting that $Y_0 = x \ne y$, $Y_{\tau_B} \notin B$, we have

$$v(y)\mu_y = \mathbb{E}^x\Big[\sum_{i=0}^{\tau_B - 1} 1_{\{Y_{i+1} = y\}}\Big] = \mathbb{E}^x\Big[\sum_{i=0}^{\tau_B - 1} \sum_z 1_{\{Y_i = z\}} P(z, y)\Big]$$

$$= \sum_z v(z)\mu_z \frac{\mu_{zy}}{\mu_z} = \sum_z v(z)\mu_{yz}$$

Dividing both sides by μ_y, we have $v(y) = \sum_z P(y, z)v(z)$, so v is harmonic on $B \setminus \{x\}$.

(ii) Since $\mathcal{E}(v, f) = f(x) = 0$ when $x \notin B$, we may assume $x \in B$. We first show

$$\mathcal{L}v(x) = -\mu_x^{-1}. \qquad (4.7)$$

Indeed, since $x \in B$, by the Markov property, we have

$$\mathbb{E}^x[L(x, \tau_B)] - 1 = \sum_y P(x, y)\mathbb{E}^y[L(x, \tau_B)].$$

Plugging $v(y) = g_B(x, y) = g_B(y, x) = \mu_x^{-1}\mathbb{E}^y[L(x, \tau_B)]$ into the above equality, we have

$$\mu_x v(x) - 1 = \mu_x \sum_y P(x, y) v(y),$$

so (4.7) is proved. Now, since v is harmonic on $B \setminus \{x\}$ and $\operatorname{Supp} f \subset B$, noting that we can apply Lemma 2.1.5 (ii) because $f \in \mathbb{L}^2$ when B is finite, we have

$$\mathcal{E}(v, f) = -\mathcal{L}v(x) f(x) \mu_x = f(x).$$

Here (4.7) is used in the last equality. We thus obtain $\mathcal{E}(v, f) = f(x)$.

(iii) Multiplying both sides of (4.4) by μ_y and summing over $y \in B$, we obtain the result.

(iv) If $x \notin B$, both sides are 0, so let $x \in B$. Let $p_B^x(z) = g_B(x, z)/g_B(x, x)$. Then, by Proposition 2.2.3 and (i) above, we see that p_B^x attains the minimum in the definition of the effective resistance. (Note that the assumption that B is finite is used to guarantee the uniqueness of the minimum.) Thus, using (ii) above,

$$R_{\text{eff}}(x, B^c)^{-1} = \mathcal{E}(p_B^x, p_B^x) = \frac{\mathcal{E}(g_B(x, \cdot), g_B(x, \cdot))}{g_B(x, x)^2} = \frac{1}{g_B(x, x)}. \qquad (4.8)$$

(v) If $x \notin B$, both sides are 0, so let $x \in B$. Using (iii), (iv) and (4.6), we have

$$\mathbb{E}^x[\tau_B] = \sum_{y \in B} g_B(x, y) \mu_y \le \sum_{y \in B} g_B(x, x) \mu_y = R_{\text{eff}}(x, B^c) \mu(B).$$

We thus obtain the desired inequality.

$\square$

Remark 4.1.2. As mentioned before Definition 2.1.3, $\mathcal{L}v(x)\mu_x$ in (4.7) is the total flux flowing into x, given the potential v. So, we see that the total flux flowing out from x is 1 when the potential $g_B(x, \cdot)$ is given at x.

The next example shows that $R_{\text{eff}}(x, B^c) = g_B(x, x)$ does not hold in general when B is not finite.

Example 4.1.3. Consider $\mathbb{Z}^3$ with weight 1 on each nearest neighbor bond. Let p be an additional point and put a bond with weight 1 between the origin of $\mathbb{Z}^3$ and p; $X = \mathbb{Z}^3 \cup \{p\}$ with the above mentioned weights is the weighted graph in this example. Let $B = \mathbb{Z}^3$ and let $B_n = B \cap B(0, n)$. By Lemma 4.1.1 (iv), $R_{\text{eff}}(0, B_n^c) = g_{B_n}(0, 0)$. If we set $c_n^{-1} := R_{\text{eff}}^{\mathbb{Z}^3}(0, B(0, n)^c)$, then it is easy to compute $R_{\text{eff}}(0, B_n^c) = (1 + c_n)^{-1}$. Since simple random walk on $\mathbb{Z}^3$ is transient, $\lim_{n \to \infty} c_n =: c_0 > 0$. As will be proved in the proof of Lemma 4.2.1, $g_B(0, 0) = \lim_{n \to \infty} g_{B_n}(0, 0)$, so $g_B(0, 0) = (1 + c_0)^{-1}$. On the other hand, it is easy to see $R_{\text{eff}}(0, B^c) = 1 > (1 + c_0)^{-1}$, so $R_{\text{eff}}(0, B^c) > g_B(0, 0)$. This also shows that in general the resistance between two sets cannot be approximated by the resistance of finite approximation graphs.

For any $A \subset X$, and $A_1, A_2 \subset X$ with $A_1 \cap A_2 = \emptyset$, $A \cap A_i = \emptyset$ (for either $i = 1$ or 2) define

$$R_{\text{eff}}^A(A_1, A_2)^{-1} = \inf\{\mathcal{E}(f, f) : f \in H^2, f|_{A_1} = 1, f|_{A_2} = 0,$$
$$f \text{ is a constant on } A\}. \tag{4.9}$$

In other word, $R_{\text{eff}}^A(\cdot, \cdot)$ is the effective resistance for the network where the set A is shorted to one point. Clearly $R_{\text{eff}}^A(x, A) = R_{\text{eff}}(x, A)$ for $x \in X \setminus A$. We then have the following (cf. [153, Theorem 4.1, 4.3]).

Proposition 4.1.4. *Let $B \subset X$ be a finite set. Then*

$$g_B(x, y) = \frac{1}{2}(R_{\text{eff}}(x, B^c) + R_{\text{eff}}(y, B^c) - R_{\text{eff}}^{B^c}(x, y)), \qquad \forall x, y \in B.$$

Proof. Since the set B^c is shorted to one point, it is enough to prove this when B^c is a point, say z. Noting that $R_{\text{eff}}^{\{z\}}(x, y) = R_{\text{eff}}(x, y)$, we will prove the following.

$$g_B(x, y) = \frac{1}{2}(R_{\text{eff}}(x, z) + R_{\text{eff}}(y, z) - R_{\text{eff}}(x, y)), \qquad \forall x, y \in B. \tag{4.10}$$

By Lemma 4.1.1 (iv) and (4.6), we have

$$g_B(x, y) = \mathbb{P}^y(\sigma_x < \tau_B)g_B(x, x) = \mathbb{P}^y(\sigma_x < \sigma_z)R_{\text{eff}}(x, z). \tag{4.11}$$

Now $V = \{x, y, z\} \subset X$ and consider the trace of the network to V, and consider the function u on V such that $u(x) = 1, u(z) = 0$ and u is harmonic on y. Using the same notation as in the proof of Lemma 4.0.1 (iv), we have

$$\mathbb{P}^y(\sigma_x < \sigma_z) = u(y) = \frac{R_{xy}^{-1}}{R_{xy}^{-1} + R_{yz}^{-1}} \times 1 + \frac{R_{yz}^{-1}}{R_{xy}^{-1} + R_{yz}^{-1}} \times 0 = \frac{R_{yz}}{R_{xy} + R_{yz}}.$$

Putting this into (4.11) and using (4.2), we have

$$g_B(x, y) = \frac{R_{yz}}{R_{xy} + R_{yz}} \cdot \frac{R_{zx}(R_{xy} + R_{yz})}{R_{xy} + R_{yz} + R_{zx}} = \frac{R_{zx}R_{yz}}{R_{xy} + R_{yz} + R_{zx}}. \tag{4.12}$$

By (4.3), we obtain (4.10). $\qquad\qquad\qquad\qquad\qquad\qquad\qquad\qquad\qquad\qquad \square$

Remark 4.1.5. Take an additional point p_0 and consider the Δ-Y transform between $V = \{x, y, z\}$ and $W = \{p_0, x, y, z\}$. Namely, $W = \{p_0, x, y, z\}$ is the network such that $\tilde{\mu}_{p_0 x}, \tilde{\mu}_{p_0 y}, \tilde{\mu}_{p_0 z} > 0$ and other weights are 0, and the trace of $(W, \tilde{\mu})$ to V is $(V, \hat{\mu})$. (See for example, [151, Lemma 2.1.15].) Then $\frac{R_{zx}R_{yz}}{R_{xy}+R_{yz}+R_{zx}}$ in (4.12) is equal to $\tilde{\mu}_{p_0 z}^{-1}$, so $g_B(x, y) = \tilde{\mu}_{p_0 z}^{-1}$.

Corollary 4.1.6. *Let $B \subset X$ be a finite set. Then*

$$|g_B(x, y) - g_B(x, z)| \leq R_{\mathrm{eff}}(y, z), \qquad \forall x, y, z \in X.$$

Proof. By Proposition 4.1.4, we have

$$|g_B(x, y) - g_B(x, z)| \leq \frac{|R_{\mathrm{eff}}(y, B^c) - R_{\mathrm{eff}}(z, B^c)| + |R_{\mathrm{eff}}^{B^c}(x, y) - R_{\mathrm{eff}}^{B^c}(x, z)|}{2}$$

$$\leq \frac{R_{\mathrm{eff}}(y, z) + R_{\mathrm{eff}}^{B^c}(y, z)}{2} \leq R_{\mathrm{eff}}(y, z),$$

which gives the desired estimate. $\square$

4.2   Green Density on a General Set

In this section, we will discuss the Green density when B can be infinite (cf. [153, Chap. 4]). Since we do not use results in this section later, readers may skip it.

Let $B \subset X$. We define the Green density by (4.4). By Proposition 2.2.9, $g_B(x, y) < \infty$ for all $x, y \in X$ when $\{Y_n\}$ is transient, whereas $g_X(x, y) = \infty$ for all $x, y \in X$ when $\{Y_n\}$ is recurrent. Since there is nothing interesting when $g_X(x, y) = \infty$, throughout this section we will only consider the case

$$B \neq X \text{ when } \{Y_n\} \text{ is recurrent.}$$

Then we can easily see that the process $\{Y_n^B\}$ killed on exiting B is transient, so $g_B(x, y) < \infty$ for all $x, y \in X$. It is easy to see that (4.5), (4.6), and Lemma 4.1.1 (i), (iii) hold without any change of the proof.

Recall that H_0^2 is the closure of $C_0(X)$ in H^2. We can generalize Lemma 4.1.1 (ii) as follows.

Lemma 4.2.1 (Reproducing Property of Green Density). *For each $x \in X$, $g_B(x, \cdot) \in H_0^2$. Further, for all $f \in H_0^2$ with $\mathrm{Supp}\, f \subset B$, it holds that $\mathcal{E}(g_B(x, \cdot), f) = f(x)$ for each $x \in X$.*

Proof. When B is finite, this is already proved in Lemma 4.1.1 (ii), so let B be infinite (and $B \neq X$ if $\{Y_n\}$ is recurrent). Fix $x_0 \in X$ and let $B_n = B(x_0, n) \cap B$ and write $v_n(z) = g_{B_n}(x, z)$. Then $\tau_{B_n} \uparrow \tau_B$ so that $v_n(z) \uparrow v(z) < \infty$ for all $z \in X$. Using the reproducing property, for $m \leq n$, we have

$$\mathcal{E}(v_n - v_m, v_n - v_m) = g_{B_n}(x, x) - g_{B_m}(x, x),$$

which implies that $\{v_n\}$ is the Cauchy sequence in H^2. It follows that $v_n \to v$ in H^2 and $v \in H_0^2$. Now for each $f \in H_0^2$ with $\mathrm{Supp}\, f \subset B$, choose $f_n \in C_0(X)$ so that

Supp $f \subset B_n$ and $f_n \to f$ in H^2. Then, as we proved above, $\mathcal{E}(f_n, v_n) = f_n(x)$. Taking $n \to \infty$ and using Lemma 2.1.2 (i), we obtain $\mathcal{E}(f, v) = f(x)$. $\square$

As we see in Example 4.1.3, $R_{\mathrm{eff}}(0, B^c) = \lim_{n \to \infty} R_{\mathrm{eff}}(0, (B \cap B(0, n)^c))$ does not hold in general. So we introduce another resistance metric as follows.

$$R_*(x, y) := \sup \left\{ \frac{|u(x) - u(y)|^2}{\mathcal{E}(u, u)} : u \in H_0^2 \oplus 1, \mathcal{E}(u, u) > 0 \right\}, \quad \forall x, y \in X, x \neq y,$$

where $H_0^2 \oplus 1 = \{f + a : f \in H_0^2, a \in \mathbb{R}\}$. Note that for this case supremum is taken over all $H_0^2 \oplus 1$, whereas for the case of the effective resistance metric, it is taken over all H^2 as in (4.1). Clearly $R_*(x, y) \leq R_{\mathrm{eff}}(x, y)$.

Here and in the following of this section, we will consider $R_*(x, B^c)$, so we assume $B \neq X$. (For $R_*(x, B^c)$, we consider B^c as one point by shorting.) Using $R_*(\cdot, \cdot)$, we can generalize Lemma 4.1.1 (iv), (v) as follows.

Lemma 4.2.2. *Let $B \subset X$ be a set such that $B \neq X$. Then the following hold.*

(i) $R_*(x, B^c) = g_B(x, x)$ *for each* $x \in X$.
(ii) $\mathbb{E}^x[\tau_B] \leq R_*(x, B^c)\mu(B)$ *for each* $x \in X$.

Proof. (i) If $x \notin B$, both sides are 0, so let $x \in B$. Let $p_B^x(z) = g_B(x, z)/g_B(x, x)$. Note that we cannot follow the proof Lemma 4.1.1 (iv) directly because we do not have uniqueness for the solution of the Dirichlet problem in general. Instead, we discuss as follows. Rewriting the definition, we have

$$R_*(x, B^c)^{-1} = \inf\{\mathcal{E}(f, f) : f \in H_{0,x}^2(B)\}, \tag{4.13}$$

where $H_{0,x}^2(B) := \{f \in H_0^2 \oplus 1 : \mathrm{Supp}\, f \subset B, f(x) = 1\}$. Take any $v \in H_{0,x}^2(B)$. (Note that $H_{0,x}^2(B) \subset H_0^2$ since $B \neq X$.) Then, by Lemma 4.2.1,

$$\mathcal{E}(v - p_B^x, p_B^x) = \frac{\mathcal{E}(v - p_B^x, g_B(x, \cdot))}{g_B(x, x)} = \frac{v(x) - p_B^x(x)}{g_B(x, x)} = 0.$$

So we have

$$\mathcal{E}(v, v) = \mathcal{E}(v - p_B^x, v - p_B^x) + \mathcal{E}(p_B^x, p_B^x) \geq \mathcal{E}(p_B^x, p_B^x),$$

which shows that the infimum in (4.13) is attained by p_B^x. Thus by (4.8) with $R_*(\cdot, \cdot)$ instead of $R_{\mathrm{eff}}(\cdot, \cdot)$, we obtain (i).

Given (i), (ii) can be proved exactly in the same way as the proof of Lemma 4.1.1 (v). $\square$

Remark 4.2.3. As mentioned in [152, Sect. 2], when (X, μ) is transient, one can show that $(\mathcal{E}, H_0^2 \oplus 1)$ is the resistance form on $X \cup \{\Delta\}$, where $\{\Delta\}$ is a new point that can be regarded as a point of infinity. Note that there is no weighted graph

$(X \cup \{\Delta\}, \bar{\mu})$ whose associated resistance form is $(\mathcal{E}, H_0^2 \oplus 1)$. Indeed, if there is, then $1_{\{\Delta\}} \in H_0^2 \oplus 1$, which contradicts the fact $1 \notin H_0^2$ (Proposition 2.2.13).

Given Lemma 4.2.2, Proposition 4.1.4 and Corollary 4.1.6 holds in general by changing $R_{\mathrm{eff}}(\cdot, \cdot)$ to $R_*(\cdot, \cdot)$ without any change of the proof.

Finally, note that by Proposition 2.2.15, we see that $H^2 = H_0^2 \oplus 1$ if and only if there is no non-constant harmonic functions of finite energy. One can see that it is also equivalent to $R_{\mathrm{eff}}(x, y) = R_*(x, y)$ for all $x, y \in X$. (The necessity can be shown by the fact that the resistance metric determines the resistance form: see [151, Sect. 2.3].)

4.3   General Heat Kernel Estimates

In this section, we give general on-diagonal upper and lower heat kernel estimates. Define

$$B(x, r) = \{y \in X : d(x, y) < r\}, \quad V(x, R) = \mu(B(x, R)).$$

For $\Omega \subset X$, let $r(\Omega)$ be the *inradius*, that is

$$r(\Omega) = \max\{r \in \mathbb{N} : \exists x_0 \in \Omega \text{ such that } B(x_0, r) \subset \Omega\}.$$

Lemma 4.3.1. *Assume that* $\inf_{x,y \in X : x \sim y} \mu_{xy} > 0$.

(i) *For any non-void finite set* $\Omega \subset X$,

$$\lambda_1(\Omega) \geq \frac{c_1}{r(\Omega)\mu(\Omega)}. \tag{4.14}$$

(ii) *Suppose that there exists a strictly increasing function* v *on* $\mathbb{N}$ *such that*

$$V(x, r) \geq v(r), \qquad \forall x \in X, \forall r \in \mathbb{N}. \tag{4.15}$$

Then, for any non-void finite set $\Omega \subset X$,

$$\lambda_1(\Omega) \geq \frac{c_1}{v^{-1}(\mu(\Omega))\mu(\Omega)}. \tag{4.16}$$

Proof. (i) Let f be any function with Supp $f \subset \Omega$ normalized as $\|f\|_\infty = 1$. Then $\|f\|_2^2 \leq \mu(\Omega)$. Now consider a point $x_0 \in X$ such that $|f(x_0)| = 1$ and the largest integer n such that $B(x_0, n) \subset \Omega$. Then $n \leq r(\Omega)$ and there exists a sequence $\{x_i\}_{i=0}^n \subset X$ such that $x_i \sim x_{i+1}$ for $i = 0, 1, \cdots, n-1$, $x_j \in \Omega$ for $j = 0, 1, \cdots, n-1$ and $x_n \notin \Omega$. So we have

$$\mathcal{E}(f,f) \geq \frac{1}{2}\sum_{i=0}^{n-1}(f(x_i)-f(x_{i+1}))^2\mu_{x_i x_{i+1}} \geq \frac{c_1}{n}\Big(\sum_{i=0}^{n-1}|f(x_i)-f(x_{i+1})|\Big)^2 \geq \frac{c_1}{n},$$

where the last inequality is due to $\sum_{i=0}^{n-1}|f(x_i) - f(x_{i+1})| \geq |f(x_0) - f(x_n)| = 1$. Combining these,

$$\frac{\mathcal{E}(f,f)}{\|f\|_2^2} \geq \frac{c_1}{n\mu(\Omega)} \geq \frac{c_1}{r(\Omega)\mu(\Omega)}.$$

Taking infimum over all such f, we obtain the result.

(ii) Denote $r = r(\Omega)$. Then, there exists $x_0 \in \Omega$ such that $B(x_0, r) \subset \Omega$, so (4.15) implies $v(r) \leq \mu(\Omega)$. Thus $r \leq v^{-1}(\mu(\Omega))$, and (4.16) follows from (4.14). □

Proposition 4.3.2 (Upper Bound: Slow Decay). *Assume that* $\inf_{x,y\in X:x\sim y}\mu_{xy} > 0$ *and*

$$V(x,r) \geq c_1 r^D, \qquad \forall x \in X, \forall r \in \mathbb{N}, \tag{4.17}$$

for some $D \geq 1$. Then the following holds.

$$\sup_{x\in X} p_t(x,x) \leq c_2 t^{-\frac{D}{D+1}}, \qquad \forall t \geq 1.$$

Proof. By (4.16), we have $\lambda_1(\Omega) \geq c_1\mu(\Omega)^{-1-1/D}$. Thus the result is obtained by Theorem 3.2.7 by taking $\theta = 2D/(D+1)$. □

Remark 4.3.3. We can generalize Proposition 4.3.2 as follows (see [26]):
Assume that $\inf_{x,y\in X:x\sim y}\mu_{xy} > 0$ and (4.15) holds for all $r \geq r_0$. Then the following holds.

$$\sup_{x\in X} p_t(x,x) \leq c_1 m(t), \qquad \forall t \geq r_0^2,$$

where m is defined by

$$t - r_0^2 = \int_{v(r_0)}^{1/m(t)} v^{-1}(s)ds.$$

Indeed, by (4.15), we see that (3.6) holds with $\varphi(s)^2 = cv^{-1}(s)s$. Thus the result can be obtained by applying Remarks 3.1.5 and 3.2.3. In fact, the above generalized version of Proposition 4.3.2 also holds for geodetically complete non-compact Riemannian manifolds with bounded geometry (see [26]).

Below is the table of the slow heat kernel decay $m(t)$, given the information of the volume growth $v(r)$.

$V(x,r) \geq$	$\exp(cr)$	$c\exp(cr^\alpha)$	cr^D	cr
$\sup_{x\in X} p_t(x,x) \leq$	$ct^{-1}\log t$	$ct^{-1}(\log t)^{1/\alpha}$	$ct^{-D/(D+1)}$	$ct^{-1/2}$

Next we discuss general form of the on-diagonal heat kernel estimate. This lower bound is quite robust, and the argument works as long as there is a Hunt process and the heat kernel exists. Note that the estimate is independent of the upper bound, so in general the two estimates may not coincide.

Proposition 4.3.4 (Lower Bound). *Let $B \subset X$ and $x \in B$. Then*

$$p_{2t}(x,x) \geq \frac{\mathbb{P}^x(\tau_B > t)^2}{\mu(B)}, \qquad \forall t > 0.$$

Proof. Using the Chapman-Kolmogorov equation and the Schwarz inequality, we have

$$\mathbb{P}^x(\tau_B > t)^2 \leq \mathbb{P}^x(Y_t \in B)^2 = \left(\int_B p_t(x,y)d\mu(y)\right)^2$$

$$\leq \mu(B)\int_B p_t(x,y)^2 d\mu(y) \leq \mu(B)p_{2t}(x,x),$$

which gives the desired inequality.   □

4.4 Strongly Recurrent Case

In this section, we will restrict ourselves to the "strongly recurrent" case and give sufficient conditions for precise on-diagonal upper and lower estimates of the heat kernel. (To be precise, Proposition 4.4.1 holds for general weighted graphs, but $F_{R,\lambda}$ in (4.23) may hold only for the strongly recurrent case.). See [27] for the definition of the strongly recurrent case.

Throughout this section, we fix a based point $0 \in X$ and let $D \geq 1, 0 < \alpha \leq 1$. As before $d(\cdot,\cdot)$ is a graph distance, but we do not use the property of the graph distance except in Remark 4.4.2. In fact all the results in this section hold for any metric (not necessarily a geodesic metric) on X without any change of the proof.

Proposition 4.4.1. *For $n \in \mathbb{N}$, let $f_n(x) = p_n(0,x) + p_{n+1}(0,x)$. Assume that $R_{\mathrm{eff}}(0,y) \leq c_* d(0,y)^\alpha$ holds for all $y \in X$. Let $r \in \mathbb{N}$ and $n = 2[r^{D+\alpha}]$. Then*

$$f_n(0) \leq c_1 n^{-\frac{D}{D+\alpha}}\left(c_* \vee \frac{r^D}{V(0,r)}\right). \tag{4.18}$$

Especially, if $c_2 r^D \leq V(0,r)$, then $f_n(0) \leq c_3 n^{-D/(D+\alpha)}$.

Proof. First, note that similarly to (2.9), we can easily check that

$$\mathcal{E}(f_n, f_n) = f_{2n}(0) - f_{2n+2}(0). \tag{4.19}$$

Choose $x_* \in B(0, r)$ such that $f_n(x_*) = \min_{x \in B(0,r)} f_n(x)$. Then

$$f_n(x_*) V(0, r) \leq \sum_{x \in B(0,r)} f_n(x) \mu_x \leq \sum_{x \in G} p_n(0, x) \mu_x + \sum_{x \in G} p_{n+1}(0, x) \mu_x \leq 2,$$

so that $f_n(x_*) \leq 2/V(0, r)$. Using Lemma 4.0.1 (iii), $R_{\text{eff}}(0, y) \leq c_* d(0, y)^\alpha$, and (4.19), we have

$$f_n(0)^2 \leq 2\left(f_n(x_*)^2 + |f_n(0) - f_n(x_*)|^2\right) \leq \frac{8}{V(0,r)^2} + 2R_{\text{eff}}(0, x_*)\mathcal{E}(f_n, f_n)$$

$$\leq \frac{8}{V(0,r)^2} + 2c_* d(0, x_*)^\alpha \mathcal{E}(f_n, f_n)$$

$$\leq \frac{8}{V(0,r)^2} + 2c_* d(0, x_*)^\alpha (f_{2n}(0) - f_{2n+2}(0)). \tag{4.20}$$

The spectral decomposition gives that $k \to f_{2k}(0) - f_{2k+2}(0)$ is non-increasing. Thus

$$n\left(f_{2n}(0) - f_{2n+2}(0)\right) \leq (2[n/2] + 1)\left(f_{4[n/2]}(0) - f_{4[n/2]+2}(0)\right)$$

$$\leq 2\sum_{i=[n/2]}^{2[n/2]} (f_{2i}(0) - f_{2i+2}(0)) \leq 2f_{2[n/2]}(0).$$

Since $n = 2[r^{D+\alpha}]$ is even, putting this into (4.20), we have $f_n(0)^2 \leq \frac{8}{V(0,r)^2} + \frac{4c_* r^\alpha f_n(0)}{n}$. Using $a + b \leq 2(a \vee b)$, we have

$$f_n(0) \leq c_1\left(\frac{1}{V(0,r)} \vee \frac{c_* r^\alpha}{n}\right). \tag{4.21}$$

Rewriting, we obtain (4.18). $\qquad\square$

Remark 4.4.2. (i) Putting $n = 2[r^\alpha V(0, r)]$ in (4.21), we have the following estimate:

$$f_{2[r^\alpha V(0,r)]}(0) \leq \frac{c_1(1 \vee c_*)}{V(0, r)}. \tag{4.22}$$

(ii) When $\alpha = 1$, using Lemma 4.0.1 (i), we see that the assumption of Proposition 4.4.1 holds if $\inf_{x,y \in X : x \sim y} \mu_{xy} > 0$. So (4.18) gives another proof of Proposition 4.3.2.

In the following, we write $B_R = B(0, R)$. Let $\varepsilon_\lambda = (3c_*\lambda)^{-1/\alpha}$. We say the weighted graph (X, μ) satisfies $F_{R,\lambda}$ (or simply say that $F_{R,\lambda}$ holds) if it satisfies the following estimates:

$$V(0, R) \leq \lambda R^D, \; R_{\text{eff}}(0, z) \leq c_* d(0, z)^\alpha, \; \forall z \in \bar{B}_R,$$

$$R_{\text{eff}}(0, B_R^c) \geq \frac{R^\alpha}{\lambda}, \; V(0, \varepsilon_\lambda R) \geq \frac{(\varepsilon_\lambda R)^D}{\lambda}, \tag{4.23}$$

where $\bar{B}_R := \{y \in X : d(0, y) \leq R\}$.

Proposition 4.4.3. *In the following, we fix $R \geq 1$ and $\lambda > 0$.*

(i) *If $V(0, R) \leq \lambda R^D$, $R_{\text{eff}}(0, z) \leq c_* d(0, z)^\alpha$ for all $z \in B_R$, then the following holds.*

$$\mathbb{E}^x[\tau_{B_R}] \leq 2c_*\lambda R^{D+\alpha} \qquad \forall x \in B_R. \tag{4.24}$$

(ii) *If $F_{R,\lambda}$ holds, then there exist $c_1 = c_1(c_*), q_0 > 0$ such that the following holds for all $x \in B(0, \varepsilon_\lambda R)$.*

$$\mathbb{E}^x[\tau_{B_R}] \geq c_1 \lambda^{-q_0} R^{D+\alpha}, \tag{4.25}$$

$$\mathbb{P}^x(\tau_{B_R} > n) \geq \frac{c_1 \lambda^{-q_0} R^{D+\alpha} - n}{2c_*\lambda R^{D+\alpha}} \quad \forall n \geq 0. \tag{4.26}$$

Proof. (i) Using Lemma 4.1.1 (v), the assumption, and the shorting law, for each $x \in B_R$ we have

$$\mathbb{E}^x[\tau_{B_R}] \leq R_{\text{eff}}(x, B_R^c)\mu(B_R) \leq (R_{\text{eff}}(0, x) + R_{\text{eff}}(0, B_R^c))\mu(B_R) \leq 2c_*\lambda R^{D+\alpha}. \tag{4.27}$$

(ii) Denote $B' = B(0, \varepsilon_\lambda R)$. By Lemma 2.2.6 and the assumption, we have for $y \in B'$,

$$\mathbb{P}^y(\tau_{B_R} < \sigma_0) = \mathbb{P}^y(\sigma_{B_R^c} < \sigma_0) \leq \frac{R_{\text{eff}}(y, B_R^c \cup \{0\})}{R_{\text{eff}}(y, B_R^c)} \leq \frac{R_{\text{eff}}(y, 0)}{R_{\text{eff}}(0, B_R^c) - R_{\text{eff}}(0, y)}$$

$$\leq \frac{c_* d(y, 0)^\alpha}{R_{\text{eff}}(0, B_R^c) - c_* d(y, 0)^\alpha} \leq \frac{R^\alpha/(3\lambda)}{R^\alpha/\lambda - R^\alpha/(3\lambda)} = \frac{1}{2}.$$

Applying this into (4.6) and using Lemma 4.1.1 (iv), we have for $y \in B'$,

$$g_{B_R}(0, y) = g_{B_R}(0, 0)\mathbb{P}^y(\sigma_0 < \tau_{B_R}) \geq \frac{1}{2} R_{\text{eff}}(0, B_R^c) \geq \frac{R^\alpha}{2\lambda}.$$

Thus,

$$\mathbb{E}^0[\tau_{B_R}] \geq \sum_{y \in B'} g_B(0, y)\mu_y \geq \frac{R^\alpha}{2\lambda}\mu(B') \geq \frac{\varepsilon_\lambda^D R^{D+\alpha}}{2\lambda^2}.$$

Further, for $x \in B'$,

$$\mathbb{E}^x[\tau_{B_R}] \geq \mathbb{P}^x(\sigma_0 < \sigma_{B_R})\mathbb{E}^0[\tau_{B_R}] \geq \frac{1}{2}\mathbb{E}^0[\tau_{B_R}] \geq \frac{\varepsilon_\lambda^D R^{D+\alpha}}{4\lambda^2},$$

so (4.25) is obtained.

Next, by (4.24), (4.25), and the Markov property of Y, we have

$$c_1 \lambda^{-q_0} R^{D+\alpha} \leq \mathbb{E}^x[\tau_{B_R}] \leq n + \mathbb{E}^x[1_{\{\tau_{B_R} > n\}}\mathbb{E}^{Y_n}[\tau_{B_R}]]$$
$$\leq n + 2c_*\lambda R^{D+\alpha}\mathbb{P}^x(\tau_{B_R} > n).$$

So (4.26) is obtained.   □

Proposition 4.4.4. *If $F_{R,\lambda}$ holds, then there exist $c_1 = c_1(c_*), q_0, q_1 > 0$ such that the following holds for all $x \in B(0, \varepsilon_\lambda R)$.*

$$p_{2n}(x, x) \geq c_1 \lambda^{-q_1} n^{-D/(D+\alpha)} \text{ for } \frac{c_{4.4.3.1}}{4\lambda^{q_0}}R^{D+\alpha} \leq n \leq \frac{c_{4.4.3.1}}{2\lambda^{q_0}}R^{D+\alpha}.$$

Proof. Using Proposition 4.3.4 and (4.26), we have

$$p_{2n}(x, x) \geq \frac{\mathbb{P}^x(\tau_{B_R} > n)^2}{\mu(B_R)} \geq \frac{(c_{4.4.3.1}/(2c_*\lambda^{q_0+1}))^2}{\lambda R^D} \geq c_1 \lambda^{-q_1} n^{-D/(D+\alpha)},$$

for some $c_1, q_1 > 0$.   □

Remark 4.4.5. The above results can be generalized as follows. Let $v, r : \mathbb{N} \to [0, \infty)$ be strictly increasing functions with $v(1) = r(1) = 1$ which satisfy

$$C_1^{-1}\left(\frac{R}{R'}\right)^{d_1} \leq \frac{v(R)}{v(R')} \leq C_1\left(\frac{R}{R'}\right)^{d_2}, \quad C_2^{-1}\left(\frac{R}{R'}\right)^{\alpha_1} \leq \frac{r(R)}{r(R')} \leq C_2\left(\frac{R}{R'}\right)^{\alpha_2}$$

for all $1 \leq R' \leq R < \infty$, where $C_1, C_2 \geq 1$, $1 \leq d_1 \leq d_2$ and $0 < \alpha_1 \leq \alpha_2 \leq 1$. Assume that the following holds instead of (4.23) for a suitably chosen $\varepsilon_\lambda > 0$:

$$V(0, R) \leq \lambda v(R), \quad R_{\text{eff}}(0, z) \leq c_* r(d(0, z)), \quad \forall z \in B_R,$$
$$R_{\text{eff}}(0, B_R^c) \geq \frac{r(R)}{\lambda}, \quad V(0, \varepsilon_\lambda R) \geq \frac{v(\varepsilon_\lambda R)}{\lambda}.$$

$x \in B(0, \varepsilon_\lambda R)$. Then, the following estimates hold.

$$\frac{c_1}{\lambda^{q_1} v(\mathcal{I}(n))} \le p_{2n}(x,x) \le \frac{c_2}{\lambda^{q_2} v(\mathcal{I}(n))}$$

for $\frac{c_0}{4\lambda^{q_0}} v(R) r(R) \le n \le \frac{c_0}{2\lambda^{q_0}} v(R) r(R)$, where $\mathcal{I}(\cdot)$ is the inverse function of $(v \cdot r)(\cdot)$ (see [166] for details).

4.5   Applications to Fractal Graphs

In this section, we will apply the estimates obtained in the previous section to fractal graphs.

Two-Dimensional Pre-Sierpinski Gasket. Let V_0 be the vertices of the pre-Sierpinski gasket (Fig. 3.1) and define $V_{-n} = 2^n V_0$. Let $a_n = (2^n, 0), b_n = (2^{n-1}, 2^{n-1}\sqrt{3})$ be the vertices in V_{-n}.

Lemma 4.5.1. *It holds that* $R_{\mathrm{eff}}(0, \{a_n, b_n\}) = \frac{1}{2}\left(\frac{5}{3}\right)^n$.

Proof. Let $p_n = \mathbb{P}^0(\sigma_{\{a_n, b_n\}} < \sigma_0^+)$. Let $z = (3/2, \sqrt{3}/2)$ and define $q_1 = \mathbb{P}^z(\sigma_{\{a_1, b_1\}} < \sigma_0^+)$. Then, by the Markov property, $p_1 = 4^{-1}(p_1 + q_1 + 1)$ and $q_1 = 2^{-1}(p_1 + 1)$. Solving them, we have $p_1 = 3/5$. Next, for a simple random walk $\{Y_k\}$ on the pre-Sierpinski gasket, define the induced random walk $\{Y_k^{(n)}\}$ on V_{-n} as follows.

$$\eta_0 = \min\{k \ge 0 : Y_k \in V_{-n}\}, \quad \eta_i = \min\{k > \eta_{i-1} : Y_k \in V_{-n} \setminus \{Y_{\eta_{i-1}}\}\},$$

$$Y_i^{(n)} = Y_{\eta_i}, \quad \text{for } i \in \mathbb{N}.$$

Then it can be easily seen that $\{Y_k^{(n)}\}$ is a simple random walk on V_{-n}. Using this fact, we can inductively show $p_{n+1} = p_n \cdot p_1 = p_n \cdot (3/5) = \cdots = (3/5)^{n+1}$. We thus obtain the result by using Theorem 2.2.5. $\qquad\square$

Let $d_f = \log 3 / \log 2$, $d_w = \log 5 / \log 2$.

Proposition 4.5.2. *The following hold for all $R \ge 1$.*

(i) $c_1 R^{d_f} \le V(0, R) \le c_2 R^{d_f}$.
(ii) $R_{\mathrm{eff}}(0, z) \le c_3 d(0, z)^{d_w - d_f}, \quad \forall z \in X$.
(iii) $R_{\mathrm{eff}}(0, B(0, R)^c) \ge c_4 R^{d_w - d_f}$.

Proof. (i) Since there are 3^n triangles with length 1 in $B(0, 2^n)$, we see that $c_1 3^n \le V(0, 2^n) \le c_2 3^n$. Next, for each R, take $n \in \mathbb{N}$ such that $2^{n-1} \le R < 2^n$. Then $c_1 3^{n-1} \le V(0, 2^{n-1}) \le V(0, R) \le V(0, 2^n) \le c_2 3^n$. Since $3 = 2^{d_f}$, we have the desired estimate.
(ii) We first prove the following:

$$c_3 (5/3)^n \le R_{\mathrm{eff}}(0, a_n) \le c_4 (5/3)^n. \tag{4.28}$$

Indeed, by the shorting law and Lemma 4.5.1, $2^{-1}(5/3)^n = R_{\text{eff}}(0, \{a_n, b_n\}) \le R_{\text{eff}}(0, a_n)$. On the other hand, let $\varphi_n^{(1)}$ and $\varphi_n^{(2)}$ be such that

$$\mathcal{E}(\varphi_n^{(1)}, \varphi_n^{(1)}) = R_{\text{eff}}(0, a_n)^{-1} = R_{\text{eff}}(0, b_n)^{-1} = \mathcal{E}(\varphi_n^{(2)}, \varphi_n^{(2)}).$$

Then, by symmetry, $\varphi_n^{(1)}(b_n) = \varphi_n^{(2)}(a_n) =: C$. So

$$\frac{1}{2}\left(\frac{5}{3}\right)^n = R_{\text{eff}}(0, \{a_n, b_n\}) \ge \mathcal{E}\left(\frac{\varphi_n^{(1)} + \varphi_n^{(2)}}{1 + C}, \frac{\varphi_n^{(1)} + \varphi_n^{(2)}}{1 + C}\right)^{-1}$$

$$\ge \left\{\frac{2}{(1+C)^2}\left(\mathcal{E}(\varphi_n^{(1)}, \varphi_n^{(1)}) + \mathcal{E}(\varphi_n^{(2)}, \varphi_n^{(2)})\right)\right\}^{-1}$$

$$= \frac{(1+C)^2}{4} R_{\text{eff}}(0, a_n).$$

We thus obtain (4.28). Now for each $z \in X$, choose $n \ge 0$ such that $2^n \le d(0, z) < 2^{n+1}$. We can then take a sequence $z = z_0, z_1, \cdots, z_n$ such that $z_i \in V_{-i}$, $d(z_i, z_{i+1}) \le 2^i$ for $i = 0, 1, \cdots, n-1$ ($z_i = z_{i+1}$ is allowed) and $d(0, z_n) = 2^n$. Similarly to (4.28), we can show that $R_{\text{eff}}(z_i, z_{i+1}) \le c_5(5/3)^i$. Thus, using the triangle inequality, we have

$$R_{\text{eff}}(0, z) \le \sum_{i=0}^{n-1} R_{\text{eff}}(z_i, z_{i+1}) + R_{\text{eff}}(0, z_n)$$

$$\le c_5\left(\sum_{i=0}^{n-1}(5/3)^i + (5/3)^n\right) \le c_6(5/3)^n \le c_7 d(0, z)^{d_w - d_f}.$$

So the desired estimate is obtained.

(iii) For each R, take $n \in \mathbb{N}$ such that $2^{n-1} \le R < 2^n$. Then, by the shorting law and Lemma 4.5.1,

$$R_{\text{eff}}(0, B(0, R)^c) \ge R_{\text{eff}}s(0, \{a_n, b_n\}) = \frac{1}{2}\left(\frac{5}{3}\right)^n = \frac{1}{2}2^{n(d_w - d_f)} \ge c_8 R^{d_w - d_f},$$

so we have the desired estimate. □

By Propositions 4.4.1, 4.4.4, 4.5.2, we see that the following heat kernel estimate holds for simple random walk on the two-dimensional Sierpinski gasket.

$$c_1 n^{-d_f/d_w} \le p_{2n}(0, 0) \le c_2 n^{-d_f/d_w}, \quad \forall n \ge 1. \tag{4.29}$$

Vicsek Sets. We next consider the Vicsek set (Fig. 4.1). Since this graph is a tree, we can apply Lemma 4.0.1 (ii), and Lemma 4.0.1 (i) trivially holds. So by a similar proof we can obtain Proposition 4.5.2 with $d_f = \log 5/\log 3$ and $d_w = d_f + 1$, so

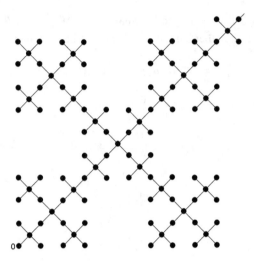

Fig. 4.1 Two-dimensional Vicsek set

(4.29) holds. This shows that the estimate in Proposition 4.3.2 is in general the best possible estimate when $D = \log 5/\log 3$. By considering the generalization of the Vicsek set in $\mathbb{R}^d$, we can see that the same is true when $D = \log(1 + 2^d)/\log 3$ (which is a sequence that goes to infinity as $d \to \infty$). In fact it is proved in [26, Theorem 5.1] that the estimate in Proposition 4.3.2 is in general the best possible estimate for any $D \geq 1$.

Remark 4.5.3. It is known that the sub-Gaussian heat kernel estimates $(UHKE(\beta))$ $+(LHKE(\beta))$ in Definition 3.3.4 hold for simple random walk on the pre-Sierpinski gasket, on Vicsek sets, and in general on the so-called nested fractal graphs with $\beta = d_w$—see for example, [120, 121, 130, 144], for books about random walks/diffusions on fractals, see [16, 151, 206, 210] etc.

Chapter 5
Heat Kernel Estimates for Random Weighted Graphs

From this chapter, we consider the situation where we have a random weighted graph $\{(X(\omega), \mu^\omega) : \omega \in \Omega\}$ on a probability space $(\Omega, \mathcal{F}, \mathbb{P})$. We assume that, for each $\omega \in \Omega$, the graph $X(\omega)$ is infinite, locally finite and connected, and contains a marked vertex $0 \in X(\omega)$. We denote balls in $X(\omega)$ by $B_\omega(x, r)$, their volume by $V_\omega(x, r)$, and write

$$B(R) = B_\omega(R) = B_\omega(0, R), \qquad V(R) = V_\omega(R) = V_\omega(0, R).$$

We write $Y = (Y_n, n \geq 0, P_\omega^x, x \in X(\omega))$ for the Markov chain on $X(\omega)$, and denote by $p_n^\omega(x, y)$ its heat kernel with respect to μ^ω. Note that from this section, we will (in principle) use $\mathbb{P}, \mathbb{E}$ for randomness of the media and use P^x, E^x for randomness of the Markov chain. Let

$$\tau_R = \tau_{B(0,R)} = \min\{n \geq 0 : Y_n \notin B(0, R)\}.$$

5.1 Random Walk on a Random Graph

Recall that $F_{R,\lambda}$ is defined in (4.23). We have the following quenched estimates (i.e. estimates which hold $\mathbb{P}$-a.s.).

Theorem 5.1.1. *Let $R_0, \lambda_0 \geq 1$. Assume that there exist $p(\lambda) \geq 0$ with $p(\lambda) \leq c_1 \lambda^{-q_0}$ for some $q_0, c_1 > 0$ such that for each $R \geq R_0$ and $\lambda \geq \lambda_0$,*

$$\mathbb{P}(\{\omega : (X(\omega), \mu^\omega) \text{ satisfies } F_{R,\lambda}\}) \geq 1 - p(\lambda). \qquad (5.1)$$

Then there exist $\alpha_1, \alpha_2 > 0$ and $\Omega_0 \subset \Omega$ with $\mathbb{P}(\Omega_0) = 1$ such that the following holds: For all $\omega \in \Omega_0$ and $x \in X(\omega)$, there exist $N_x(\omega), R_x(\omega) \in \mathbb{N}$ such that

T. Kumagai, *Random Walks on Disordered Media and their Scaling Limits*, Lecture Notes in Mathematics 2101, DOI 10.1007/978-3-319-03152-1_5,
© Springer International Publishing Switzerland 2014

$$(\log n)^{-\alpha_1} n^{-\frac{D}{D+\alpha}} \le p^\omega_{2n}(x,x) \le (\log n)^{\alpha_1} n^{-\frac{D}{D+\alpha}}, \quad \forall n \ge N_x(\omega), \qquad (5.2)$$

$$(\log R)^{-\alpha_2} R^{D+\alpha} \le E^x_\omega \tau_R \le (\log R)^{\alpha_2} R^{D+\alpha}, \quad \forall R \ge R_x(\omega). \qquad (5.3)$$

Further, if (5.1) holds with $p(\lambda) \le \exp(-c_2\lambda^{q_0})$ for some $q_0, c_2 > 0$, then (5.2), (5.3) hold with $\log\log n$ instead of $\log n$.

Proof. We will take $\Omega_0 = \Omega_1 \cap \Omega_2$ where Ω_1 and Ω_2 are defined below.

First we prove (5.2). Write $w(n) = p^\omega_{2n}(0,0)$. By Propositions 4.4.1 and 4.4.4, we have, taking $n = [c_1(\lambda)R^{D+\alpha}]$,

$$\mathbb{P}((c_1\lambda^{q_1})^{-1} \le n^{D/(D+\alpha)} w(n) \le c_1\lambda^{q_1}) \ge 1 - 2p(\lambda). \qquad (5.4)$$

Let $n_k = [e^k]$ and $\lambda_k = k^{2/q_0}$ (by choosing R suitably). Then, since $\sum p(\lambda_k) < \infty$, by the Borel–Cantelli lemma there exists $K_0(\omega)$ with $\mathbb{P}(K_0 < \infty) = 1$ such that $c_1^{-1} k^{-2q_1/q_0} \le n_k^{D/(D+\alpha)} w(n_k) \le c_1 k^{2q_1/q_0}$ for all $k \ge K_0(\omega)$. Let $\Omega_1 = \{K_0 < \infty\}$. For $k \ge K_0$ we therefore have

$$c_2^{-1}(\log n_k)^{-2q_1/q_0} n_k^{-D/(D+\alpha)} \le w(n_k) \le c_2(\log n_k)^{2q_1/q_0} n_k^{-D/(D+\alpha)},$$

so that (5.2) holds for the subsequence n_k. The spectral decomposition gives that $p^\omega_{2n}(0,0)$ is monotone decreasing in n. So, if $n > N_0 = e^{K_0} + 1$, let $k \ge K_0$ be such that $n_k \le n < n_{k+1}$. Then

$$w(n) \le w(n_k) \le c_2(\log n_k)^{2q_1/q_0} n_k^{-D/(D+\alpha)} \le 2e^{D/(D+\alpha)} c_2(\log n)^{2q_1/q_0} n^{-D/(D+\alpha)}.$$

Similarly $w(n) \ge w(n_{k+1}) \ge c_3 n^{-D/(D+\alpha)}(\log n)^{-2q_1/q_0}$. Taking $q_2 > 2q_1/q_0$, so that the constants c_2, c_3 can be absorbed into the $\log n$ term, we obtain (5.2) for $x = 0$.

If $x, y \in X(\omega)$ and $k = d_\omega(x,y)$, then using the Chapman–Kolmogorov equation

$$p^\omega_{2n}(x,x)(p^\omega_k(x,y)\mu_x(\omega))^2 \le p^\omega_{2n+2k}(y,y).$$

Let $\omega \in \Omega_1$, $x \in X(\omega)$, write $k = d_\omega(0,x)$, $h^\omega(0,x) = (p^\omega_k(x,0)\mu_x(\omega))^{-2}$, and let $n \ge N_0(\omega) + 2k$. Then

$$p^\omega_{2n}(x,x) \le h^\omega(0,x) p^\omega_{2n+2k}(0,0) \le h^\omega(0,x) \frac{(\log(n+k))^{q_2}}{(n+k)^{D/(D+\alpha)}}$$

$$\le h^\omega(0,x) \frac{(\log(2n))^{q_2}}{n^{D/(D+\alpha)}} \le \frac{(\log n)^{1+q_2}}{n^{D/(D+\alpha)}}$$

provided $\log n \ge 2^{q_2} h^\omega(0,x)$. Taking

$$N_x(\omega) = \exp(2^{q_2} h^\omega(0,x)) + 2d_\omega(0,x) + N_0(\omega), \qquad (5.5)$$

and $\alpha_1 = 1 + q_2$, this gives the upper bound in (5.2). The lower bound is obtained in the same way.

Next we prove (5.3). Write $F(R) = E^x_\omega \tau_R$. By (4.24) and (4.25), we have

$$\mathbb{P}(c_1 \lambda^{-q_0} R^{D+\alpha} \leq F(R) \leq 2c_2 \lambda R^{D+\alpha}, \ \forall x \in B(0, \varepsilon_\lambda R)) \geq 1 - p(\lambda). \quad (5.6)$$

Let $R_n = e^n$ and $\lambda_n = n^{2/q_0}$, and let F_n be the event of the left hand side of (5.6) when $R = R_n$, $\lambda = \lambda_n$. Then we have $\mathbb{P}(F_n^c) \leq p(\lambda_n) \leq n^{-2}$, so by the Borel–Cantelli, if $\Omega_2 = \liminf F_n$, then $\mathbb{P}(\Omega_2) = 1$. Hence there exist M_0 with $M_0(\omega) < \infty$ on Ω_2, and $c_3, q_3 > 0$ such that for $\omega \in \Omega_2$ and $x \in X(\omega)$,

$$(c_3 \lambda_n^{q_3})^{-1} \leq \frac{F(R_n)}{R_n^{D+\alpha}} \leq c_3 \lambda_n^{q_3}, \quad (5.7)$$

provided $n \geq M_0(\omega)$ and n is also large enough so that $x \in B(\varepsilon_{\lambda_n} R_n)$. Writing $M_x(\omega)$ for the smallest such n,

$$c_3^{-1} (\log R_n)^{-2q_3/q_0} R_n^{D+\alpha} \leq F(R_n) \leq c_3 (\log R_n)^{2q_3/q_0} R_n^{D+\alpha}, \ \forall n \geq M_x(\omega).$$

As $F(R)$ is monotone increasing, the same argument as in the proof of (5.2) above enables us to replace $F(R_n)$ by $F(R)$, for all $R \geq R_x = 1 + e^{M_x}$. Taking $\alpha_2 > 2q_3/q_0$, we obtain (5.3).

The case $p(\lambda) \leq \exp(-c_2 \lambda^{q_0})$ can be proved similarly by the following changes; take $\lambda_k = (e + (2/c_2) \log k)^{1/q_0}$ instead of $\lambda_k = k^{2/q_0}$, and take $N_x(\omega) = \exp(\exp(C h^\omega(0, x))) + 2d_\omega(0, x) + N_0(\omega)$ in (5.5). Then, $\log n$ (resp. $\log n_k$, $\log R_n$) in the above proof are changed to $\log \log n$ (resp. $\log \log n_k$, $\log \log R_n$) and the proof goes through.   □

We also have the following annealed (averaged) estimates. Note that there are no log terms in them.

Proposition 5.1.2. *Suppose (5.1) holds for some $p(\lambda) \geq 0$ which goes to 0 as $\lambda \to \infty$, and suppose the following hold,*

$$\mathbb{E}[R_{\mathrm{eff}}(0, B(R)^c) V(R)] \leq c_1 R^{D+\alpha}, \qquad \forall R \geq 1. \quad (5.8)$$

Then

$$c_2 R^{D+\alpha} \leq \mathbb{E}[E^0_\omega \tau_R] \leq c_3 R^{D+\alpha}, \quad \forall R \geq 1, \quad (5.9)$$

$$c_4 n^{-D/(D+\alpha)} \leq \mathbb{E}[p^\omega_{2n}(0,0)], \quad \forall n \geq 1. \quad (5.10)$$

Suppose in addition that there exist $c_5 > 0$, $\lambda_0 > 1$ and $q_0' > 2$ such that

$$\mathbb{P}(\lambda^{-1} R^D \leq V(R), \ R_{\mathrm{eff}}(0, y) \leq \lambda d(0, y)^\alpha, \ \forall y \in B(R)) \geq 1 - \frac{c_5}{\lambda^{q_0'}}, \quad (5.11)$$

for each $R \geq 1, \lambda \geq \lambda_0$. Then

$$\mathbb{E}[p_{2n}^\omega(0,0)] \leq c_6 n^{-D/(D+\alpha)}, \qquad \forall n \geq 1. \tag{5.12}$$

Proof. We begin with the upper bounds in (5.9). By (4.27) and (5.8),

$$\mathbb{E}[E_\omega^0 \tau_R] \leq \mathbb{E}[R_{\text{eff}}(0, B(R)^c)V(R)] \leq c_1 R^{D+\alpha}.$$

For the lower bounds, it is sufficient to find a set $F \subset \Omega$ of "good" graphs with $\mathbb{P}(F) \geq c_2 > 0$ such that, for all $\omega \in F$ we have suitable lower bounds on $E_\omega^0 \tau_R$ or $p_{2n}^\omega(0,0)$. We assume that $R \geq 1$ is large enough so that $\varepsilon_{\lambda_0} R \geq 1$, where λ_0 is chosen large enough so that $p(\lambda_0) < 1/4$. We can then obtain the results for all n (chosen below to depend on R) and R by adjusting the constants c_2, c_4 in (5.9) and (5.10).

Let F be the event of the left hand side of (5.6) when $\lambda = \lambda_0$. Then $\mathbb{P}(F) \geq \frac{3}{4}$, and for $\omega \in F$, $E_\omega^0 \tau_R \geq c_3 \lambda_0^{-q_0} R^{D+\alpha}$. So,

$$\mathbb{E}[E_\cdot^0 \tau_R] \geq \mathbb{E}[E_\cdot^0 \tau_R : F] \geq c_3 \lambda_0^{-q_0} R^{D+\alpha} \mathbb{P}(F) \geq \frac{3c_3}{4} \lambda_0^{-q_0} R^{D+\alpha}.$$

Also, by (5.4), if $n = [c_4(\lambda_0)R^{D+\alpha}]$, then $p_{2n}^\omega(0,0) \geq c_5 \lambda_0^{-q_1} n^{-D/(D+\alpha)}$. So, given $n \in \mathbb{N}$, choose R so that $n = [c_4(\lambda_0)R^{D+\alpha}]$ and let F be the event of the left hand side of (5.4). Then $\mathbb{P}(F) \geq \frac{1}{2}$, and

$$\mathbb{E}p_{2n}^\cdot(0,0) \geq \mathbb{P}(F)c_5 \lambda_0^{-q_1} n^{-D/(D+\alpha)} \geq \frac{c_5}{2} \lambda_0^{-q_1} n^{-D/(D+\alpha)},$$

giving the lower bound in (5.10).

Finally we prove (5.12). Let H_k be the event of the left hand side of (5.11) with $\lambda = k$. By (4.18), we see that $p_{2n}^\omega(0,0) \leq c_6 k n^{-D/(D+\alpha)}$ if $\omega \in H_k$, where R is chosen to satisfy $n = 2[R^{D+\alpha}]$. Since $\mathbb{P}(\cup_k H_k) = 1$, using (5.11), we have

$$\mathbb{E}p_{2n}^\omega(0,0) \leq \sum_k c_6(k+1)n^{-D/(D+\alpha)}\mathbb{P}(H_{k+1} \setminus H_k)$$

$$\leq \sum_k c_6(k+1)n^{-D/(D+\alpha)}\mathbb{P}(H_k^c) \leq c_7 n^{-D/(D+\alpha)} \sum_k (k+1)k^{-q_0'} < \infty,$$

since $q_0' > 2$. We thus obtain (5.12). $\qquad\qquad\qquad\qquad\qquad\qquad\qquad\square$

Remark 5.1.3. (i) With some extra efforts, one can obtain quenched estimates of $\max_{0 \leq k \leq n} d(0, Y_k)$ and $\mu(W_n)$ where $W_n = \{Y_0, Y_1, \cdots, Y_n\}$ (range of the random walk), and annealed lower bound of $E_\omega^0 d(0, Y_n)$. See [166, (1.23), (1.28), (1.31)] and [32, (1.16), (1.20), (1.23)].

(ii) In [80], fluctuations in the heat kernel estimates are studied in the resistance form, assuming the fluctuation of the volume growth.

5.2 The IIC and the Alexander–Orbach Conjecture

The problem of random walk on a percolation cluster—the "ant in the labyrinth" [110]—has received much attention both in the physics and the mathematics literature. From the next chapter, we will consider random walk on a percolation cluster, so $X(\omega)$ will be a percolation cluster.

Let us first recall the bond percolation model on the lattice $\mathbb{Z}^d$ with $\{\{x, y\} : x, y \in \mathbb{Z}^d, \|x - y\| = 1\}$ being the set of bonds (edges) where $\| \cdot \|$ is the Euclidean norm. Each bond is open with probability $p \in (0, 1)$, independently of all the others. Let $\mathcal{C}(x)$ be the open cluster (i.e. the open connected component) containing x; then if $\theta(p) = P_p(|\mathcal{C}(x)| = +\infty)$ it is well known (see [122]) that for $d \geq 2$, there exists $p_c = p_c(d) \in (0, 1)$ such that $\theta(p) = 0$ if $p < p_c$ and $\theta(p) > 0$ if $p > p_c$.

If $d = 2$ or $d \geq 19$ ($d > 6$ for spread-out models mentioned below) it is known that $\theta(p_c) = 0$, and it is conjectured that this holds for $d \geq 2$—see for example, [122, 203]. (Quite recently, [102] improves the lower bound $d \geq 19$ using a new way of lace expansion; see [101].)

At the critical probability $p = p_c$, it is known (to some extent) that in any box of side n there exist with high probability open clusters of diameter of order n (see [148, Sect. 5.1]). For large n the local properties of these large finite clusters can, in certain circumstances, be captured by regarding them as subsets of an infinite cluster $\mathcal{G}$, called the incipient infinite cluster (IIC for short). This IIC $\mathcal{G} = \mathcal{G}(\omega)$ is our random weighted graph $X(\omega)$.

IIC was constructed when $d = 2$ in [149], by taking the limit as $N \to \infty$ of the cluster $\mathcal{C}(0)$ conditioned to intersect the boundary of a box of side N with center at the origin. For large d a construction of the IIC in $\mathbb{Z}^d$ is given in [138], using the lace expansion. It is believed that the results there will hold for any $d > 6$. Reference [138] also gives the existence and some properties of the IIC for all $d > 6$ for spread-out models: these include the case when there is a bond between x and y with probability pL^{-d} whenever y is in a cube side L with center x, and the parameter L is large enough. We write $\mathcal{G}_d$ for the IIC in $\mathbb{Z}^d$. It is believed that the global properties of $\mathcal{G}_d$ are the same for all $d > d_c$, both for nearest neighbor and spread-out models. (Here d_c is the *critical dimension* which is 6 for the percolation model.) In [138] it is proved for spread-out models that any two infinite paths in $\mathcal{G}_d$ intersect infinitely often. In particular, $\mathcal{G}_d$ has a single end (i.e., the complement of every finite subgraph has a unique infinite connected component). See [203] for a summary of the high-dimensional results.

Let $Y = \{Y_n^\omega\}_{n \in \mathbb{N}}$ be the simple random walk on $\mathcal{G}_d = \mathcal{G}_d(\omega)$, and $p_n^\omega(x, y)$ be its heat kernel. Define the *spectral dimension* of $\mathcal{G}_d$ by

$$d_s(\mathcal{G}_d) = -2 \lim_{n \to \infty} \frac{\log p_{2n}^\omega(x, x)}{\log n},$$

(if this limit exists). Alexander and Orbach [8] conjectured that, for any $d \geq 2$, $d_s(\mathcal{G}_d) = 4/3$. While it is now thought that this is unlikely to be true for small d

(see [140, Sect. 7.4]), the results on the geometry of $\mathcal{G}_d$ for spread-out models in [138] are consistent with this holding for d above the critical dimension.

Recently, it is proved that the Alexander–Orbach conjecture holds for random walk for the IIC on a tree [33], on a high dimensional oriented percolation cluster [32], and on a high dimensional percolation cluster [162]. In all cases, we apply Theorem 5.1.1, namely we verify (5.1) with $D = 2, \alpha = 1$ to prove the Alexander–Orbach conjecture. We will discuss details in the next two chapters.

Chapter 6
Alexander–Orbach Conjecture Holds When Two-Point Functions Behave Nicely

This chapter is based on the paper by Kozma-Nachmias [162] with some simplification by [197]. Our framework here is unimodular transitive graphs that contains $\mathbb{Z}^d$ as a typical example.

6.1 The Model and Main Theorems

We write $x \leftrightarrow y$ if x is connected to y by an open path. We write $x \overset{r}{\leftrightarrow} y$ if there is an open path of length less than or equal to r that connects x and y.

Definition 6.1.1. Let (X, E) be an infinite graph.

(i) A bijection map $f : X \to X$ is called a graph automorphism for X if $\{f(u), f(v)\} \in E$ if and only if $\{u, v\} \in E$. Denote the set of all the automorphisms of X by $\mathrm{Aut}(X)$

(ii) (X, E) is said to be transitive if for any $u, v \in X$, there exists $\phi \in \mathrm{Aut}(X)$ such that $\phi(u) = v$.

(iii) For each $x \in X$, define the stabilizer of x by

$$S(x) = \{\phi \in \mathrm{Aut}(X) : \phi(x) = x\}.$$

(iv) A transitive graph X is unimodular if for each $x, y \in X$,

$$|\{\phi(y) : \phi \in S(x)\}| = |\{\phi(x) : \phi \in S(y)\}|.$$

A typical example of unimodular transitive graphs is $\mathbb{Z}^d$ with nearest neighbor bonds. Indeed, if we define $\psi_x : \mathbb{Z}^d \to \mathbb{Z}^d$ ($x \in \mathbb{Z}^d$) by $\psi_x(z) = z + x$, then $\psi_x \in \mathrm{Aut}(\mathbb{Z}^d)$. $\mathbb{Z}^d$ is transitive because for each $u, v \in \mathbb{Z}^d$, $\psi_{v-u}(u) = v$. For each $x, y \in \mathbb{Z}^d$, define a map $F_{xy} : S(x) \to S(y)$ by $F_{xy}(\phi) = \psi_{y-x} \circ \phi \circ \psi_{x-y}$. It is easy to see that this is a bijective map, so $\mathbb{Z}^d$ is unimodular.

T. Kumagai, *Random Walks on Disordered Media and their Scaling Limits*, Lecture Notes in Mathematics 2101, DOI 10.1007/978-3-319-03152-1_6,
© Springer International Publishing Switzerland 2014

Let (X, μ) be a unimodular transitive weighted graph with weight 1 on each bond, i.e. $\mu_{xy} = 1$ for $x \sim y$. Let $d_X(\cdot, \cdot)$ be the graph distance on X.

We fix a root $0 \in X$ as before. (Note that since the graph is transitive, the results in this chapter is independent of the choice of $0 \in X$.) Consider a bond percolation on X and let $p_c = p_c(X)$ be its critical probability, namely

$$\mathbb{P}_p(\text{there exists an open } \infty\text{-cluster}) > 0 \quad \text{for} \quad p > p_c(X),$$

$$\mathbb{P}_p(\text{there exists an open } \infty\text{-cluster}) = 0 \quad \text{for} \quad p < p_c(X).$$

In this chapter, we will consider the case $p = p_c$ and denote $\mathbb{P} := \mathbb{P}_{p_c}$. Throughout this chapter, we assume that the following limit exists (independently on how $d_X(0, x) \to \infty$).

$$\mathbb{P}_{\mathrm{IIC}}(F) = \lim_{d_X(0,x) \to \infty} \mathbb{P}(F \,|\, 0 \leftrightarrow x) \tag{6.1}$$

for any cylinder event F (i.e., an event that depends only on the status of a finite number of edges). We denote the realization of IIC by $\mathcal{G}(\omega)$. For the critical bond percolations on $\mathbb{Z}^d$, this is proved in [138] for d large using the lace expansion.

We consider the following two conditions. The first is the triangle condition by Aizenman-Newman [4], which indicates mean-field behavior of the model (see also [3]):

$$\sum_{x,y \in X} \mathbb{P}(0 \leftrightarrow x)\mathbb{P}(x \leftrightarrow y)\mathbb{P}(y \leftrightarrow 0) < \infty. \tag{6.2}$$

Note that the value of the left hand side of (6.2) does not change if 0 is changed to any $v \in X$ because X is unimodular and transitive.

The second is the following condition for two-point functions: There exist $c_0, c_1, c_2 > 0$ and a decreasing function $\psi : \mathbb{R}_+ \to \mathbb{R}_+$ with $\psi(r) \leq c_0 \psi(2r)$ for all $r > 0$ such that

$$c_1 \psi(d_X(0, x)) \leq \mathbb{P}(0 \leftrightarrow x) \leq c_2 \psi(d_X(0, x)) \qquad \forall x \in X, \tag{6.3}$$

where $d_X(\cdot, \cdot)$ is the original graph distance on X. Because X is transitive, for any $u, v \in X$, there exists $\phi_{uv} \in \mathrm{Aut}(X)$ such that $\phi_{uv}(u) = v$. It can be easily checked that $d_X(x, y) = d_X(0, \phi_{x0}(y))$ and $\mathbb{P}(x \leftrightarrow y) = \mathbb{P}(0 \leftrightarrow \phi_{x0}(y))$, so (6.3) gives control of all two-point functions.

Since X is unimodular and transitive, we can deduce the following from the above mentioned two conditions.

Lemma 6.1.2. *(i) Equation (6.2) implies the following open triangle condition:*

$$\lim_{K \to \infty} \sup_{w:d_X(0,w) \geq K} \sum_{x,y \in X} \mathbb{P}(0 \leftrightarrow x)\mathbb{P}(x \leftrightarrow y)\mathbb{P}(y \leftrightarrow w) = 0. \tag{6.4}$$

(ii) Equation (6.4) implies the following estimates: There exist $C_1, C_2, C_3 > 0$ such that

$$\mathbb{P}\big(|\mathcal{C}(0)| > n\big) \le C_1 n^{-1/2}, \qquad \forall n \ge 1, \qquad (6.5)$$

$$C_2(p_c - p)^{-1} \le \mathbb{E}[|\mathcal{C}(0)|] \le C_3(p_c - p)^{-1}, \qquad \forall p < p_c. \qquad (6.6)$$

where $\mathcal{C}(0)$ is the open connected component containing 0.

Proof. (i) This is proved in [161] (see [37, Lemma 2.1] for $\mathbb{Z}^d$).
(ii) Equation (6.4) implies (6.6) (see [4, Proposition 3.2] for $\mathbb{Z}^d$ and [198, p. 291] for general unimodular transitive graphs), and the following for $h > 0$ small (see [37, (1.13)] for $\mathbb{Z}^d$ and [198, p. 292] for general unimodular transitive graphs):

$$c_1 h^{1/2} \le \sum_{j=1}^{\infty} \mathbb{P}(|\mathcal{C}(0)| = j)(1 - e^{-jh}) \le c_2 h^{1/2}.$$

Taking $h = 1/n$, we have $\mathbb{P}\big(|\mathcal{C}(0)| > n\big) \le C_1 n^{-1/2}$. $\qquad\square$

For critical bond percolations on $\mathbb{Z}^d$, (6.3) with $\psi(x) = x^{2-d}$ was obtained by Hara et al. [134] for the spread-out model and $d > 6$, and by Hara [133] for the nearest-neighbor model with $d \ge 19$ using the lace expansion. (They obtained the right asymptotic behavior including the constant.) Given (6.3) with $\psi(x) = x^{2-d}$ for $d > 6$, it is easy to check that (6.2) holds as well.

For each subgraph G of X, write G_p for the result of p-bond percolation. Allowing some abuse of notation, we denote

$$B(x, r; G) = \{z : d_{G_{p_c}}(x, z) \le r\}, \quad \partial B(x, r; G) = \{z : d_{G_{p_c}}(x, z) = r\}$$

where $d_{G_{p_c}}(x, z)$ is the length of the shortest path between x and z in G_{p_c} (it is ∞ if there is no such path), and $p_c = p_c(X)$. We write

$$B(x, r) := B(x, r; X),$$

in other word $B(x, r)$ is a ball in C_∞ with respect to the graph distance of the cluster. Note that this notation applies only in this chapter; $B(x, r)$ in this chapter corresponds to $B(x, r + 1)$ in other chapters. Now define

$$H(r; G) = \{\partial B(0, r; G) \ne \emptyset\}, \quad \Gamma(r) = \sup_{G \subset X} \mathbb{P}(H(r; G)).$$

Note that

$$\Gamma(r) = \sup_{G \subset X} \mathbb{P}(\{\partial B(v, r; G) \ne \emptyset\}), \quad \mathbb{E}[|B(0, r)|] = \mathbb{E}[|B(v, r)|], \qquad \forall v \in X,$$

$$(6.7)$$

since X is transitive. The following two propositions play a key role.

Proposition 6.1.3. *Assume that the triangle condition (6.2) holds. Then there exists a constant $C > 0$ such that the following hold for all $r \geq 1$.*

$$\text{i) } \mathbb{E}[|B(0,r)|] \leq Cr, \quad \text{ii) } \Gamma(r) \leq C/r.$$

Proposition 6.1.4. *Assume that (6.3) and i), ii) in Proposition 6.1.3 hold. Then (5.1) in Theorem 5.1.1 holds for $\mathbb{P}_{\text{IIC}}$ with $p(\lambda) = \lambda^{-1/2}$, $D = 2$ and $\alpha = 1$.*

Consider simple random walk on the IIC and let $p_n^{\omega}(\cdot, \cdot)$ be its heat kernel. Combining the above two propositions with Theorem 5.1.1, we can derive the following theorem.

Theorem 6.1.5. *Assume that (6.2) and (6.3) hold. Then there exist $\alpha_1, \alpha_2 > 0$ and $N_0(\omega), R_0(\omega) \in \mathbb{N}$ with $\mathbb{P}(N_0(\omega) < \infty) = \mathbb{P}(R_0(\omega) < \infty) = 1$ such that*

$$(\log n)^{-\alpha_1} n^{-\frac{2}{3}} \leq p_{2n}^{\omega}(0,0) \leq (\log n)^{\alpha_1} n^{-\frac{2}{3}}, \qquad \forall n \geq N_0(\omega), \ \mathbb{P}_{\text{IIC}} - a.e. \ \omega,$$

$$(\log R)^{-\alpha_2} R^3 \leq E_{\omega}^0 \tau_R \leq (\log R)^{\alpha_2} R^3, \qquad \forall R \geq R_0(\omega), \ \mathbb{P}_{\text{IIC}} - a.e. \ \omega,$$

where τ_R is the exit time from the ball of radius R centered at 0 with respect to the graph distance of the IIC. In particular, the Alexander–Orbach conjecture holds for the IIC.

By the above mentioned reason, for the critical bond percolations on $\mathbb{Z}^d$, the Alexander–Orbach conjecture holds for the IIC for the spread-out model with $d > 6$, and for the nearest-neighbor model with $d \geq 19$.

Remark 6.1.6. (i) In fact, the existence of the limit in (6.1) is not relevant in the arguments. Indeed, even if the limit does not exist, subsequential limits exist due to compactness, and the above results hold for each limit. So the conclusions of Theorem 6.1.5 hold for any IIC measure (i.e. any subsequential limit).

(ii) The opposite inequalities of Theorem 6.1.3 (i.e. $\mathbb{E}[|B(0,r)|] \geq C'r$ and $\Gamma(r) \geq C'/r$ for all $r \geq 1$) hold under weaker assumption. See Proposition 6.3.2.

Remark 6.1.7. (i) Note that in this section the graph distance of the IIC is used for the estimates in Theorem 6.1.5. In [136], asymptotic behavior of the exit times for the random walk on IIC (both nearest neighbor and spread-out percolations in high dimensions) are obtained in terms of the Euclidean distance. In the paper, they also consider the critical long-range percolation, and obtain asymptotic behavior of random walk on the backbone of the IIC when d is large. The estimates are given both in terms of the Euclidean distance and the IIC graph distance.

(ii) Concerning the long-range percolation, heat kernel estimates and scaling limits are considered for the supercritical long-range percolation model and its variants in [73, 78, 79]. We will not pursue the long-range case here.

In the following sections, we prove Propositions 6.1.3 and 6.1.4.

6.2 Proof of Proposition 6.1.4

The proof splits into three lemmas.

Lemma 6.2.1. *Assume that (6.3) and i) in Proposition 6.1.3 hold. Then there exists a constant $C > 0$ such that for $r \geq 1$ and $x \in X$ with $d_X(0, x) \geq 2r$, we have*

$$\mathbb{P}(|B(0, r)| \geq \lambda r^2 | 0 \leftrightarrow x) \leq C\lambda^{-1}, \quad \forall \lambda \geq 1. \tag{6.8}$$

Proof. It is enough to prove the following for $r \geq 1$ and $x \in X$ with $d_X(0, x) \geq 2r$:

$$\mathbb{E}[|B(0, r)| \cdot 1_{\{0 \leftrightarrow x\}}] \leq c_1 r^2 \psi(d_X(0, x)). \tag{6.9}$$

Indeed, we then have, using (6.3) and (6.9),

$$\mathbb{P}(|B(0, r)| \geq \lambda r^2 | 0 \leftrightarrow x) \leq \frac{\mathbb{E}[|B(0, r)| | 0 \leftrightarrow x]}{\lambda r^2}$$

$$= \frac{\mathbb{E}[|B(0, r)| \cdot 1_{\{0 \leftrightarrow x\}}]}{\lambda r^2 \mathbb{P}(0 \leftrightarrow x)} \leq \frac{c_1 r^2 \psi(d_X(0, x))}{\lambda r^2 c_2 \psi(d_X(0, x))} \leq c_3 \lambda^{-1},$$

which gives (6.8). So we will prove (6.9).

We have

$$\mathbb{E}[|B(0, r)| \cdot 1_{\{0 \leftrightarrow x\}}] = \sum_z \mathbb{P}(0 \overset{r}{\leftrightarrow} z, 0 \leftrightarrow x) \leq \sum_{z,y} \mathbb{P}(\{0 \overset{r}{\leftrightarrow} y\} \circ \{y \overset{r}{\leftrightarrow} z\} \circ \{y \leftrightarrow x\})$$

$$\leq \sum_{z,y} \mathbb{P}(0 \overset{r}{\leftrightarrow} y) \mathbb{P}(y \overset{r}{\leftrightarrow} z) \mathbb{P}(y \leftrightarrow x). \tag{6.10}$$

Here the first inequality is because, if $\{0 \overset{r}{\leftrightarrow} z, 0 \leftrightarrow x\}$ occurs, then there must exist some y such that $\{0 \overset{r}{\leftrightarrow} y\}$, $\{y \overset{r}{\leftrightarrow} z\}$ and $\{y \leftrightarrow x\}$ occur disjointly. The second inequality uses the BK inequality twice. Recall that the BK inequality implies that for increasing events A and B that depend on only finitely many edges (such events are called local events) we have $\mathbb{P}(A \circ B) \leq \mathbb{P}(A)\mathbb{P}(B)$, where $A \circ B$ denotes disjoint occurrence (see for example, [122]). Note that $\{y \leftrightarrow x\}$ is not a local event, but it can be approximated by local events so the BK inequality still applies.

For $d_X(0, y) \leq r$ and $d_X(0, x) \geq 2r$, we have $d_X(x, y) \geq d_X(0, x) - d_X(0, y) \geq d_X(0, x)/2$, so that

$$\mathbb{P}(y \leftrightarrow x) \leq c_4 \psi(d_X(x, y)) \leq c_4 \psi(d_X(0, x)/2) \leq c_5 \psi(d_X(0, x)).$$

Thus,

$$(\text{RHS of } (6.10)) \leq c_5 \psi(d_X(0,x)) \sum_{z,y} \mathbb{P}(0 \overset{r}{\leftrightarrow} y)\mathbb{P}(y \overset{r}{\leftrightarrow} z)$$

$$\leq c_5 r \psi(d_X(0,x)) \sum_{y} \mathbb{P}(0 \overset{r}{\leftrightarrow} y) \leq c_5 r^2 \psi(d_X(0,x)).$$

Here we use i) in Proposition 6.1.3 to sum, first over z (note that (6.7) is used here) and then over y in the second and the third inequality. We thus obtain (6.9). □

Lemma 6.2.2. *Assume that (6.3) and ii) in Proposition 6.1.3 hold. Then there exists a constant $C > 0$ such that for $r \geq 1$ and $x \in X$ with $d_X(0,x) \geq 2r$, we have*

$$\mathbb{P}(|B(0,r)| \leq r^2\lambda^{-1}|0 \leftrightarrow x) \leq C\lambda^{-1}, \quad \forall \lambda \geq 1. \tag{6.11}$$

Proof. It is enough to prove the following for $r \geq 1$, $\varepsilon < 1$ and $x \in X$ with $d_X(0,x) \geq 2r$:

$$\mathbb{P}\Big(|B(0,r)| \leq \varepsilon r^2, 0 \leftrightarrow x\Big) \leq c_1 \varepsilon \psi(d_X(0,x)). \tag{6.12}$$

Indeed, we then have, using (6.3) and (6.12),

$$\mathbb{P}(|B(0,r)| \leq r^2\lambda^{-1}|0 \leftrightarrow x) = \frac{\mathbb{P}(|B(0,r)| \leq r^2\lambda^{-1}, 0 \leftrightarrow x)}{\mathbb{P}(0 \leftrightarrow x)}$$

$$\leq \frac{c_1\lambda^{-1}\psi(d_X(0,x))}{c_2\psi(d_X(0,x))} \leq c_3\lambda^{-1},$$

which gives (6.11). So we will prove (6.12).

If $|B(0,r)| \leq \varepsilon r^2$, there must exist $j \in [r/2, r]$ such that $|\partial B(0,j)| \leq 2\varepsilon r$. We fix the smallest such j. Now, if $0 \leftrightarrow x$, there exists a vertex $y \in \partial B(0,j)$ which is connected to x by a path that does not use any of the vertices in $B(0, j-1)$. We say this "$x \leftrightarrow y$ off $B(0, j-1)$". Let A be a subgraph of X such that $\mathbb{P}(B(0,j) = A) > 0$. It is clear that, for any A and any $y \in \partial A$, $\{y \leftrightarrow x$ off $A \setminus \partial A\}$ is independent of $\{B(0,j) = A\}$, where ∂A is the set of vertices in A furthest from 0 for the graph distance of A. Thus,

$$\mathbb{P}(0 \leftrightarrow x \mid B(0,j) = A) \leq \sum_{y \in \partial A} \mathbb{P}(y \leftrightarrow x \text{ off } A \setminus \partial A \mid B(0,j) = A)$$

$$= \sum_{y \in \partial A} \mathbb{P}(y \leftrightarrow x \text{ off } A \setminus \partial A) \leq \sum_{y \in \partial A} \mathbb{P}(y \leftrightarrow x)$$

$$\leq C|\partial A|\psi(d_X(0,x)), \tag{6.13}$$

where we used $d_X(x,y) \geq d_X(0,x) - d_X(0,y) \geq d_X(0,x)/2$ in the last inequality. By the definition of j we have $|\partial A| \leq 2\varepsilon r$ and summing over all A with $\mathbb{P}(B(0,j) = A) > 0$ and $\partial B(0, r/2) \neq \emptyset$ (because $0 \leftrightarrow x$) gives

$$\mathbb{P}(|B(0,r)| \leq \varepsilon r^2, 0 \leftrightarrow x) \leq C\varepsilon r \psi(d_X(0,x)) \cdot \sum_A \mathbb{P}(B(0,j) = A).$$

Since the events $\{B(0,j) = A_1\}$ and $\{B(0,j) = A_2\}$ are disjoint for $A_1 \neq A_2$, we have

$$\sum_A \mathbb{P}(B(0,j) = A) \leq \mathbb{P}(\partial B(0,r/2) \neq \emptyset) \leq c/r,$$

where ii) in Proposition 6.1.3 is used in the last inequality. We thus obtain (6.12).

□

Lemma 6.2.3. *Assume that (6.3) and i), ii) in Proposition 6.1.3 hold. Then there exists a constant $C > 0$ such that for $r \geq 1$ and $x \in X$ with $d_X(0,x) \geq 4r$, we have*

$$\mathbb{P}(R_{\text{eff}}(0, \partial B(0,r)) \leq r\lambda^{-1}|0 \leftrightarrow x) \leq C\lambda^{-1/2}, \quad \forall \lambda \geq 1. \tag{6.14}$$

Proof of Proposition 6.1.4. Note that $R_{\text{eff}}(0,z) \leq d(0,z) := d_{X_{p_c}}(0,z)$ (d is the graph distance in the critical percolation cluster) holds for all $z \in B(0,R)$, and $|B(0,R)| \leq V(0,R) \leq c_1|B(0,R)|$ for all $R \geq 1$. (We used the fact that the graph has controlled weights here.) By Lemmas 6.2.1–6.2.3, for each $R \geq 1$ and $x \in X$ with $d_X(0,x) \geq 4R$, we have

$$\mathbb{P}(\{\omega : B(0,R) \text{ satisfies } F_{R,\lambda}\}|0 \leftrightarrow x) \geq 1 - 3C\lambda^{-1/2} \quad \forall \lambda \geq 1.$$

Using (6.1), we obtain the desired estimate.   □

So, all we need is to prove Lemma 6.2.3. We first give definition of *lane* introduced in [186] (similar notion was also given in [33]).

Definition 6.2.4. (i) An edge e between $\partial B(0,j-1)$ and $\partial B(0,j)$ is called a lane for r if there exists a path with initial edge e from $\partial B(0,j-1)$ to $\partial B(0,r)$ that does not return to $\partial B(0,j-1)$.
(ii) Let $0 < j < r$ and $\lambda \geq 1$. We say that a level j has λ-lanes for r if there exist at least λ edges between $\partial B(0,j-1)$ and $\partial B(0,j)$ which are lanes for r.
(iii) We say that 0 is λ-lane rich for r if more than half of $j \in [r/4, r/2]$ have λ-lanes for r.

Note that if 0 is *not* λ-lane rich for r, then

$$R_{\text{eff}}(0, \partial B(0,r)) \geq \frac{r}{8\lambda}. \tag{6.15}$$

Indeed, since 0 is not λ-lane rich, there exist $j_1, j_2, \cdots, j_l \in [r/4, r/2]$, $l \geq r/8$ that do not have λ-lanes. For $j \in [r/4, r/2]$, let

$$J_j = \{e : e \text{ is a lane for } r \text{ that is between } \partial B(0,j-1) \text{ and } \partial B(0,j)\}. \tag{6.16}$$

Then $\{J_{j_k}\}_{k=1}^l$ are disjoint cut-sets separating 0 from $\partial B(0, r)$. Since $|J_{j_k}| \leq \lambda$, by the shorting law we have $R_{\text{eff}}(0, \partial B(0, r)) \geq \sum_{k=1}^l |J_{j_k}|^{-1} \geq l/\lambda \geq r/(8\lambda)$, so that (6.15) holds.

Proof of Lemma 6.2.3. It is enough to prove the following for $r \geq 1$, $\lambda > 1$ and $x \in X$ with $d_X(0, x) \geq 4r$:

$$\mathbb{P}\left(R_{\text{eff}}(0, \partial B(0, r)) \leq \lambda^{-1} r, \ 0 \leftrightarrow x\right) \leq c_1 \lambda^{-1/2} \psi(d_X(0, x)). \qquad (6.17)$$

Indeed, we then have, using (6.3) and (6.17),

$$\mathbb{P}(R_{\text{eff}}(0, \partial B(0, r)) \leq r\lambda^{-1} | 0 \leftrightarrow x) = \frac{\mathbb{P}(R_{\text{eff}}(0, \partial B(0, r)) \leq r\lambda^{-1}, 0 \leftrightarrow x)}{\mathbb{P}(0 \leftrightarrow x)}$$

$$\leq \frac{c_1 \lambda^{-1/2} \psi(d_X(0, x))}{c_2 \psi(d_X(0, x))} \leq c_3 \lambda^{-1/2},$$

which gives (6.14). So we will prove (6.17). The proof consists of two steps.

Step 1 We will prove the following; There exists a constant $C > 0$ such that for any $r \geq 1$, for any event E measurable with respect to $B(0, r)$ and for any $x \in X$ with $d_X(0, x) \geq 4r$,

$$\mathbb{P}(E \cap \{0 \leftrightarrow x\}) \leq C \sqrt{r \mathbb{P}(E)} \psi(d_X(0, x)). \qquad (6.18)$$

We first note that by (6.9), there exists some $j \in [r, 2r]$ such that

$$\mathbb{E}\left[|\partial B(0, j)| \cdot 1_{\{0 \leftrightarrow x\}}\right] \leq Cr\psi(d_X(0, x)).$$

Using this and the Chebyshev inequality, for each $M > 0$ we have

$$\mathbb{P}(E \cap \{0 \leftrightarrow x\}) \leq \mathbb{P}(|\partial B(0, j)| > M, 0 \leftrightarrow x) + \mathbb{P}(E \cap \{|\partial B(0, j)| \leq M, 0 \leftrightarrow x\})$$

$$\leq \frac{Cr\psi(d_X(0, x))}{M} + \mathbb{P}(E \cap \{|\partial B(0, j)| \leq M, 0 \leftrightarrow x\}).$$

For the second term, similarly to (6.13) we have, for each A,

$$\mathbb{P}(\{B(0, j) = A\} \cap \{0 \leftrightarrow x\}) \leq C |\partial A| \psi(d_X(0, x)) \mathbb{P}(B(0, j) = A).$$

Summing over all subgraphs A which satisfy E (measurability of E with respect to $B(0, r)$ is used here) and $|\partial A| \leq M$ gives $\mathbb{P}(E \cap \{|\partial B(0, j)| \leq M, 0 \leftrightarrow x\}) \leq CM\psi(d_X(0, x))\mathbb{P}(E)$. Thus

$$\mathbb{P}(E \cap \{0 \leftrightarrow x\}) \leq \frac{Cr\psi(d_X(0, x))}{M} + CM\psi(d_X(0, x))\mathbb{P}(E).$$

Taking $M = \sqrt{r/\mathbb{P}(E)}$, we obtain (6.18).

Step 2 For $j \in [r/4, r/2]$, let us condition on $B(0, j)$, take an edge e between $\partial B(0, j-1)$ and $\partial B(0, j)$, and denote the end vertex of e in $\partial B(0, j)$ by v_e. Let G_j be a graph that one gets by removing all the edges with at least one vertex in $B(0, j-1)$. Then, $\{e$ is a lane for $r\} \subset \{\partial B(v_e, r/2; G_j) \neq \emptyset\}$ in the graph G_j. By the definition of Γ and ii) in Proposition 6.1.3 (note that (6.7) is used here), we have

$$\mathbb{P}(\partial B(v_e, r/2; G_j) \neq \emptyset | B(0, j)) \leq \Gamma(r/2) \leq C/r.$$

Recall the definition of J_j in (6.16). By the above argument, we obtain

$$\mathbb{E}[|J_j| \mid B(0, j)] = \mathbb{E}(\sum_e 1_{\{e \text{ is a lane for } r\}} | B(0, j))$$

$$\leq \mathbb{E}(\sum_e 1_{\{\partial B(v_e, r/2; G_j) \neq \emptyset\}} | B(0, j)) \leq \frac{C}{r} |\partial B(0, j)|,$$

where the sum is taken over all edges between $\partial B(0, j-1)$ and $\partial B(0, j)$. This together with i) in Proposition 6.1.3 implies

$$\mathbb{E}[\sum_{j=r/4}^{r/2} |J_j|] \leq \frac{C}{r} \sum_{j=r/4}^{r/2} \mathbb{E}[|\partial B(0, j)|] \leq \frac{C}{r} \mathbb{E}[|B(0, r)|] \leq C'.$$

So, $\mathbb{P}(0$ is λ-lane rich for $r) \leq C/(\lambda r)$. Combining this with (6.15), we obtain

$$\mathbb{P}\left(R_{\text{eff}}(0, \partial B(0, r)) \leq \lambda^{-1} r\right) \leq \frac{C}{\lambda r}.$$

(Here $R_{\text{eff}}(0, \partial B(0, r)) = \infty$ if $\partial B(0, r) = \emptyset$.) This together with (6.18) in Step 1 implies (6.17).   □

6.3 Proof of Proposition 6.1.3 i)

The original proof of Proposition 6.1.3 i) by Kozma-Nachmias [162] (with some simplification in [160]) used an induction scheme which is new and nice, but it requires several pages. There is now a nice short proof by Sapozhnikov [197] which we will follow. The proof is also robust in the sense we do not need the graph to be transitive (nor unimodular).

Proposition 6.3.1. *If there exists $C_1 > 0$ such that $\mathbb{E}_p[|C(0)|] < C_1(p_c - p)^{-1}$ for all $p < p_c$, then there exists $C_2 > 0$ such that $\mathbb{E}[|B(0, r)|] \leq C_2 r$ for all $r \geq 1$.*

Proof. It is sufficient to prove for $r \geq 2/p_c$. For $p < p_c$, we will consider the coupling of percolation with parameter p and p_c as follows. First, each edge is open with probability p_c and closed with $1 - p_c$ independently. Then, for each open edge, the edge is kept open with probability p/p_c and gets closed with $1 - p/p_c$ independently. By the construction, for each $r \in \mathbb{N}$, we have

$$\mathbb{P}_p(x \overset{r}{\leftrightarrow} y) \geq \left(\frac{p}{p_c}\right)^r \mathbb{P}_{p_c}(x \overset{r}{\leftrightarrow} y), \quad \forall x, y \in X.$$

Summing over $y \in X$ and using $\mathbb{P}_p(x \overset{r}{\leftrightarrow} y) \leq \mathbb{P}_p(x \leftrightarrow y)$ and the assumption, we have

$$\mathbb{E}[|B(0,r)|] \leq \left(\frac{p_c}{p}\right)^r \mathbb{E}_p[|\mathcal{C}(0)|] \leq \left(\frac{p_c}{p}\right)^r (p_c - p)^{-1}.$$

Taking $p = p_c - 1/r$, we obtain the result.   □

Using Lemma 6.1.2, we see that (6.2) implies (6.6) for unimodular transitive graphs. So the proof of Proposition 6.1.3 i) is completed.

As mentioned in Remark 6.1.6, the opposite inequalities of Proposition 6.1.3 hold under weaker assumption. Let X be a connected, locally finite graph with $0 \in X$.

Proposition 6.3.2. *(i) If X is transitive, then there exists $c_1 > 0$ such that $\mathbb{E}[|B(0,r)|] \geq c_1 r$ for all $r \geq 1$.*

(ii) If X is unimodular transitive and satisfies (6.2), then there exists $c_2 > 0$ such that $\Gamma(r) \geq c_2/r$ for all $r \geq 1$.

Proof. (i) It is enough to prove $\mathbb{E}[|\{x : d(0,x) = r\}|] \geq 1$ for $r \geq 1$. (Note the difference of d and d_X.) Assume by contradiction that $\mathbb{E}(|\{x : d(0,x) = r_0\}|) \leq 1 - c$ for some $r_0 \in \mathbb{N}$ and $c > 0$. Let $G(r) = \mathbb{E}(|\{x : d(0,x) \leq r\}|)$. Then

$$G(2r_0) - G(r_0) = \mathbb{E}[|\{x : d(0,x) \in (r_0, 2r_0]\}|]$$
$$\leq \mathbb{E}[|\{(y,x) : d(0,y) = r_0, d(y,x) \in (1, r_0], y \text{ is on a geodesic from } 0 \text{ to } x\}|]$$
$$\leq \mathbb{E}[|\{y : d(0,y) = r_0\}|] \cdot G(r_0) \leq (1-c)G(r_0).$$

where we used Reimer's inequality and the transitivity of X in the second inequality. (Note that we can use Reimer's inequality here because $H := \{y : d(0,y) = r_0\}$ can be verified by examining all the open edges of $B(0, r_0)$ and the closed edges in its boundary, while $\{x : d(y,x) \leq r_0\}$ for $y \in H$ can be verified using open edges outside $B(0, r_0)$.) Similarly $G(nr_0) \leq (1-c)G((n-1)r_0) + G(r_0)$. A simple calculation shows that this implies that $G(nr_0) \not\to \infty$ as $n \to \infty$, which contradicts the fact that $\mathbb{E}[|\mathcal{C}(0)|] = \infty$.

(ii) We use a second moment argument. By (i) and Proposition 6.1.3 i), we have $\mathbb{E}|B(0, \lambda r)| \geq c_1 \lambda r$ and $\mathbb{E}|B(0, r)| \leq c_2 r$ for each $r \geq 1$, $\lambda \geq 1$. Putting $\lambda = 2c_2/c_1$, we get

$$\mathbb{E}|B(0, \lambda r) \setminus B(0, r)| \geq c_1 \lambda r - c_2 r = c_2 r.$$

Next, noting that $\{0 \overset{\lambda r}{\leftrightarrow} x, 0 \overset{\lambda r}{\leftrightarrow} y\} \subset \{0 \overset{\lambda r}{\leftrightarrow} z\} \circ \{z \overset{\lambda r}{\leftrightarrow} x\} \circ \{z \overset{\lambda r}{\leftrightarrow} y\}$ for some $z \in X$, the BK inequality gives

$$\mathbb{E}[|B(0, \lambda r)|^2] \leq \sum_{x,y,z} \mathbb{P}(0 \overset{\lambda r}{\leftrightarrow} z)\mathbb{P}(z \overset{\lambda r}{\leftrightarrow} x)\mathbb{P}(z \overset{\lambda r}{\leftrightarrow} y) = \left[\sum_{x \in \mathbb{Z}^d} \mathbb{P}(0 \overset{\lambda r}{\leftrightarrow} x) \right]^3 \leq c_3 r^3,$$

where the last inequality is due to Proposition 6.1.3 i). The "inverse Chebyshev" inequality (the Paley-Zygmund inequality) $\mathbb{P}(Z > 0) \geq (\mathbb{E}Z)^2/\mathbb{E}Z^2$, valid for any non-negative random variable Z yields that

$$\mathbb{P}\big(|B(0, \lambda r) \setminus B(0, r)| > 0\big) \geq \frac{c_2^2 r^2}{c_3 r^3} \geq \frac{c_4}{r},$$

which completes the proof since $\{|B(0, \lambda r) \setminus B(0, r)| > 0\} \subset H(r)$.   □

Note that the idea behind the above proof of (i) is that if $\mathbb{E}[|\{x : d(0, x) = r\}|] < 1$ then the percolation process is dominated above by a subcritical branching process which has finite mean, and this contradicts the fact that the mean is infinity. This argument works not only for the boundary of balls for the graph distance, but also for the boundary of balls for any reasonable geodesic metric. I learned the above proof of (i), which is based on that of [163, Lemma 3.1], from G. Kozma and A. Nachmias (Personal communications). The proof of (ii) is from that of [162, Theorem 1.3 (ii)].

6.4   Proof of Proposition 6.1.3 ii)

In order to prove Proposition 6.1.3 ii), we will only need (6.5). First, note that (6.5) implies the following estimate: There exists $C > 0$ such that

$$\mathbb{P}\big(|\mathcal{C}_G(0)| > n\big) \leq \frac{C_1}{n^{1/2}} \qquad \forall G \subset X, \forall n \geq 1, \tag{6.19}$$

because $|\mathcal{C}(0)| \geq |\mathcal{C}_G(0)|$. Here $\mathcal{C}_G(0)$ is the connected component containing 0 for G_{p_c}, where $p_c = p_c(X)$.

The key idea of the proof of Proposition 6.1.3 ii) is to make a regeneration argument, which is similar to the one given in the proof of Lemma 6.2.3.

Proof of Proposition 6.1.3 ii). Let $A \geq 1$ be a large number that satisfies $3^3 A^{2/3} + C_1 A^{2/3} \leq A$, where C_1 is from (6.19). We will prove that $\Gamma(r) \leq 3Ar^{-1}$. For this, it suffices to prove

$$\Gamma(3^k) \leq \frac{A}{3^k}, \tag{6.20}$$

for all $k \in \mathbb{N} \cup \{0\}$. Indeed, for any r, by choosing k such that $3^{k-1} \leq r < 3^k$, we have

$$\Gamma(r) \leq \Gamma(3^{k-1}) \leq \frac{A}{3^{k-1}} < \frac{3A}{r}.$$

We will show (6.20) by induction—it is trivial for $k = 0$ since $A \geq 1$. Assume that (6.20) holds for all $j < k$ and we prove it for k. Let $\varepsilon > 0$ be a (small) constant to be chosen later. For any $G \subset X$, we have

$$\mathbb{P}(H(3^k; G)) \leq \mathbb{P}\left(\partial B(0, 3^k; G) \neq \emptyset, |\mathcal{C}_G(0)| \leq \varepsilon 9^k\right) + \mathbb{P}\left(|\mathcal{C}_G(0)| > \varepsilon 9^k\right)$$

$$\leq \mathbb{P}\left(\partial B(0, 3^k; G) \neq \emptyset, |\mathcal{C}_G(0)| \leq \varepsilon 9^k\right) + \frac{C_1}{\sqrt{\varepsilon 3^k}}, \tag{6.21}$$

where the last inequality is due to (6.19). We claim that

$$\mathbb{P}\left(\partial B(0, 3^k; G) \neq \emptyset, |\mathcal{C}_G(0)| \leq \varepsilon 9^k\right) \leq \varepsilon 3^{k+1}(\Gamma(3^{k-1}))^2. \tag{6.22}$$

If (6.22) holds, then by (6.21) and the induction hypothesis, we have

$$\mathbb{P}(H(3^k; G)) \leq \varepsilon 3^{k+1}(\Gamma(3^{k-1}))^2 + \frac{C_1}{\sqrt{\varepsilon 3^k}} \leq \frac{\varepsilon 3^3 A^2 + C_1 \varepsilon^{-1/2}}{3^k}. \tag{6.23}$$

Put $\varepsilon = A^{-4/3}$. Since (6.23) holds for any $G \subset X$, we have

$$\Gamma(3^k) \leq \frac{3^3 A^{2/3} + C_1 A^{2/3}}{3^k} \leq \frac{A}{3^k},$$

where the last inequality is by the choice of A. This completes the inductive proof of (6.20).

So, we will prove (6.22). Observe that if $|\mathcal{C}_G(0)| \leq \varepsilon 9^k$ then there exists $j \in [3^{k-1}, 2 \cdot 3^{k-1}]$ such that $|\partial B(0, j; G)| \leq \varepsilon 3^{k+1}$. We fix the smallest such j. If, in addition, $\partial B(0, 3^k; G) \neq \emptyset$ then at least one vertex y of the $\varepsilon 3^{k+1}$ vertices of level j satisfies $\partial B(y, 3^{k-1}; G_2) \neq \emptyset$, where $G_2 \subset G$ is determined from G by removing all edges needed to calculate $B(0, j; G)$. By (6.7) and definition of Γ (with G_2), this has probability $\leq \Gamma(3^{k-1})$. Summarizing, we have

$$\mathbb{P}\Big(\partial B(0, 3^k; G) \neq \emptyset, \, |\mathcal{C}_G(0)| \leq \varepsilon 9^k \, \Big| \, B(0, j; G)\Big) \leq \varepsilon 3^{k+1} \Gamma(3^{k-1}).$$

We now sum over possible values of $B(0, j; G)$ and get an extra term of $\mathbb{P}(H(3^{k-1}; G))$ because we need to reach level 3^{k-1} to begin with (from 0). Since $\mathbb{P}(H(3^{k-1}; G)) \leq \Gamma(3^{k-1})$, we obtain (6.22). $\qquad\square$

Chapter 7
Further Results for Random Walk on IIC

In this chapter, we will summarize results for random walks on various IICs. (Some of the models discussed in this chapter are not directly related to IIC, though.)

7.1 Random Walk on IIC for Critical Percolation on Trees

Let $n_0 \geq 2$ and let $\mathbb{B}$ be the n_0-ary homogeneous tree with a root 0. We consider the critical bond percolation on $\mathbb{B}$, i.e. let $\{\eta_e : e$ is a bond on $\mathbb{B}\}$ be i.i.d. such that $P(\eta_e = 1) = \frac{1}{n_0}$, $P(\eta_e = 0) = 1 - \frac{1}{n_0}$ for each e. Set

$$\mathcal{C}(0) = \{x \in \mathbb{B} : \exists \eta\text{-open path from 0 to } x\}.$$

Let $\mathbb{B}_n = \{x \in \mathbb{B} : d(0, x) = n\}$ where $d(\cdot, \cdot)$ is the graph distance. Then, it is easy to see that $Z_n := |\mathcal{C}(0) \cap \mathbb{B}_n|$ is a branching process with offspring distribution $\text{Bin}(n_0, \frac{1}{n_0})$. Since $E[Z_1] = 1$, $\{Z_n\}$ dies out with probability 1, so $\mathcal{C}(0)$ is a finite cluster P-a.s. In this case, we can construct the incipient infinite cluster easily as follows.

Lemma 7.1.1 (Kesten [150]). *Let* $A \subset \mathbb{B}_{\leq k} := \{x \in \mathbb{B} : d(0, x) \leq k\}$. *Then*

$$\exists \lim_{n \to \infty} P(\mathcal{C}(0) \cap \mathbb{B}_{\leq k} = A | Z_n \neq 0) = |A \cap \mathbb{B}_k| P(\mathcal{C}(0) \cap \mathbb{B}_{\leq k} = A) =: \mathbb{P}_{\text{IIC}}(A).$$

Further, there exists a unique probability measure $\mathbb{P}$ *which is an extension of* $\mathbb{P}_{\text{IIC}}$ *to a probability on the set of* ∞*-connected subsets of* $\mathbb{B}$ *containing 0.*

Let $(\Omega, \mathcal{F}, \mathbb{P})$ be the probability space given above and for each $\omega \in \Omega$, let $\mathcal{G}(\omega)$ be the rooted labeled tree with distribution $\mathbb{P}$. So, $(\Omega, \mathcal{F}, \mathbb{P})$ governs the randomness of the media and for each $\omega \in \Omega$, $\mathcal{G}(\omega)$ is the incipient infinite cluster (IIC) on $\mathbb{B}$.

T. Kumagai, *Random Walks on Disordered Media and their Scaling Limits*, Lecture Notes in Mathematics 2101, DOI 10.1007/978-3-319-03152-1_7,
© Springer International Publishing Switzerland 2014

For each $\mathcal{G} = \mathcal{G}(\omega)$, let $\{Y_n\}$ be a simple random walk on $\mathcal{G}$. Let μ be a measure on $\mathcal{G}$ given by $\mu(A) = \sum_{x \in A} \mu_x$, where μ_x is the number of open bonds connected to $x \in \mathcal{G}$. Define $p_n^\omega(x, y) := \mathbb{P}^x(Y_n = y)/\mu_y$.

In this example, (5.1) in Theorem 5.1.1 holds for $\mathbb{P}$ with $p(\lambda) = \exp(-c\lambda)$, $D = 2$ and $\alpha = 1$, so we can obtain the following (7.1). In [33], further results are obtained for this example.

Theorem 7.1.2. *(i) There exist $c_0, c_1, c_2, \alpha_1 > 0$ and a positive random variable $S(x)$ with $\mathbb{P}(S(x) \geq m) \leq \frac{c_0}{\log m}$ for all $x \in \mathbb{B}$ such that the following holds for all $n \geq S(x), x \in \mathbb{B}$,*

$$c_1 n^{-2/3} (\log \log n)^{-\alpha_1} \leq p_{2n}^\omega(x, x) \leq c_2 n^{-2/3} (\log \log n)^{\alpha_1}. \qquad (7.1)$$

(ii) There exists $C \in (0, \infty)$ such that for each $\varepsilon \in (0, 1)$, the following holds for $\mathbb{P}$-a.e. ω

$$\liminf_{n \to \infty} (\log \log n)^{(1-\varepsilon)/3} n^{2/3} p_{2n}^\omega(0, 0) \leq C. \qquad (7.2)$$

We will give sketch of the proof in the next section.

Equation (7.2) together with (5.10) show that one cannot take $\alpha_1 = 0$ in (7.1). Namely, there is a oscillation of order log log for the quenched heat kernel estimates. Note that this model also satisfies the assumption given in Proposition 5.1.2, so there is no oscillation (up to constants multiples) for the annealed (averaged) heat kernel estimates.

Remark 7.1.3. (i) For $N \in \mathbb{N}$, let $\tilde{Z}_n^{(N)} = N^{-1/3} d(0, Y_{Nn})$, $n \geq 0$. In [150], Kesten proved that $\mathbb{P}$-distribution of $\tilde{Z}_n^{(N)}$ converges as $N \to \infty$. Especially, $\{\tilde{Z}^{(N)}\}$ is tight with respect to the annealed (averaged) law $\mathbb{P}^* := \mathbb{P} \times P_\omega^0$. On the other hand, by (7.2), it can be shown that $\{\tilde{Z}^{(N)}\}$ is NOT tight with respect to the quenched law (see [33]).

(ii) In [81], it is proved that the $\mathbb{P}$-distribution of the rescaled simple random walk on IIC converges to Brownian motion on the Aldous tree [5].

(iii) Equation (5.2) is proved with $D = 2$ and $\alpha = 1$ for simple random walk on IIC of the family tree for the critical Galton-Watson branching process with finite variance offspring distribution [106], and for invasion percolation on regular trees [13].

(iv) The behavior of simple random walk on IIC is different for the family tree of the critical Galton-Watson branching process with infinite variance offspring distribution. In [87], it is proved that when the offspring distribution is in the domain of attraction of a stable law with index $\beta \in (1, 2)$, then (5.2) holds with $D = \beta/(\beta - 1)$ and $\alpha = 1$. In particular, the spectral dimension of the random walk is $2\beta/(2\beta - 1)$. It is further proved that there is an oscillation of order log for the quenched heat kernel estimates. Namely, it is proved that there exists $\alpha_2 > 0$ such that

$$\liminf_{n\to\infty} n^{\frac{\beta}{2\beta-1}} (\log n)^{\alpha_2} p_{2n}^{\omega}(0,0) = 0, \quad \mathbb{P}\text{-a.e. } \omega.$$

Note that for this case convergence of $\mathbb{P}$-distribution of $N^{-\frac{\beta-1}{2\beta-1}} d(0, Y_{Nn})$ as $N \to \infty$ is already proved in [150].

One may wonder if the off-diagonal heat kernel estimates enjoy sub-Gaussian estimates given in Definition 3.3.4. As we have seen, there is an oscillation already in the on-diagonal estimates for quenched heat kernel, so one cannot expect estimates precisely the same one as in Definition 3.3.4. However, the following theorem shows that such sub-Gaussian estimates hold with high probability for the quenched heat kernel, and the precise sub-Gaussian estimates hold for annealed (averaged) heat kernel.

Let $\{Y_t\}_{t\geq 0}$ be the continuous time Markov chain with based measure μ. Set $q_t^{\omega}(x, y) = \mathbb{P}^x(Y_t = y)/\mu_y$, and let $\mathbb{P}_{x,y}(\cdot) = \mathbb{P}(\cdot|x, y \in \mathcal{G})$, $\mathbb{E}_{x,y}(\cdot) = \mathbb{E}(\cdot|x, y \in \mathcal{G})$. Then the following holds.

Theorem 7.1.4 ([33]).

(i) *Quenched heat kernel bounds:*

(a) *Let $x, y \in \mathcal{G} = \mathcal{G}(\omega)$, $t > 0$ be such that $N := [\sqrt{d(x, y)^3/t}] \geq 8$ and $t \geq c_1 d(x, y)$. Then, there exists $F_* = F_*(x, y, t)$ with $\mathbb{P}_{x_0, y_0}(F_*(x, y, t)) \geq 1 - c_2 \exp(-c_3 N)$, such that*

$$q_t^{\omega}(x, y) \leq c_4 t^{-2/3} \exp(-c_5 N), \qquad \text{for all } \omega \in F_*.$$

(b) *Let $x, y \in \mathcal{G}$ with $x \neq y$, $m \geq 1$ and $\kappa \geq 1$. Then there exists $G_* = G_*(x, y, m, \kappa)$ with $\mathbb{P}_{x_0, y_0}(G_*(x, y, m, \kappa)) \geq 1 - c_6 \kappa^{-1}$ such that*

$$q_{2T}^{\omega}(x, y) \geq c_7 T^{-2/3} e^{-c_8(\kappa+c_9)m}, \text{ for all } \omega \in G_* \text{ where } T = d(x, y)^3 \kappa/m^2.$$

(ii) *Annealed heat kernel bounds: Let $x, y \in \mathbb{B}$. Then*

$$c_1 t^{-2/3} \exp\left(-c_2 \left(\frac{d(x, y)^3}{t}\right)^{1/2}\right) \leq \mathbb{E}_{x,y} q_t^{\cdot}(x, y)$$

$$\leq c_3 t^{-2/3} \exp\left(-c_4 \left(\frac{d(x, y)^3}{t}\right)^{1/2}\right), \quad (7.3)$$

where the upper bound is for $c_5 d(x, y) \leq t$ and the lower bound is for $c_5(d(x, y) \vee 1) \leq t$.

Note that (7.3) coincides the estimate in Definition 3.3.4 with $\beta = 3$ and $V(x, r) \asymp r^2$.

7.2   Sketch of the Proof of Theorem 7.1.2

For $x \in \mathcal{G}$ and $r \geq 1$, let $M(x, r)$ be the smallest number $m \in \mathbb{N}$ such that there exists $A = \{z_1, \ldots, z_m\}$ with $d(x, z_i) \in [r/4, 3r/4]$, $1 \leq i \leq m$, so that any path γ from x to $B(x, r)^c$ must pass through the set A. Similarly to (6.15), we have

$$R_{\text{eff}}(0, B(0, R)^c) \geq R/(2M(0, R)). \tag{7.4}$$

So, in order to prove (5.1) with $p(\lambda) = \exp(-c\lambda)$ (which implies Theorem 7.1.2 (i) due to the last assertion of Theorem 5.1.1), it is enough to prove the following.

Proposition 7.2.1.   *(i) Let $\lambda > 0$, $r \geq 1$. Then*

$$\mathbb{P}(V(0, r) > \lambda r^2) \leq c_0 \exp(-c_1 \lambda), \quad \mathbb{P}(V(0, r) < \lambda r^2) \leq c_2 \exp(-c_3/\sqrt{\lambda}).$$

(ii) Let $r, m \geq 1$. Then

$$\mathbb{P}(M(x, r) \geq m) \leq c_4 e^{-c_5 m}.$$

Proof (Sketch). Basically, all the estimates can be obtained through large deviation estimates of the total population size of the critical branching process.

(i) Since $\mathcal{G}$ is a tree, $|B(x, r)| \leq V(x, r) \leq 2|B(x, r)|$, so we consider $|B(x, r)|$. Let η be the offspring distribution and $p_i := P(\eta = i)$. (Note that $E[\eta] = 1$, $\text{Var}[\eta] < \infty$.) Now define the size biased branching process $\{\tilde{Z}_n\}$ as follows: $\tilde{Z}_0 = 1$, $P(\tilde{Z}_1 = j) = (j + 1)p_{j+1}$ for all $j \geq 0$, and $\{\tilde{Z}_n\}_{n \geq 2}$ is the usual branching process with offspring distribution η. Let $\tilde{Y}_n := \sum_{k=0}^{n} \tilde{Z}_k$. Then,

$$\tilde{Y}_{r/2}[r/2] \overset{(d)}{\leq} |B(0, r)| \overset{(d)}{\leq} \tilde{Y}_r[r].$$

Here, for random variable ξ, we denote $\xi[n] \overset{(d)}{:=} \sum_{i=1}^{n} \xi_i$, where $\{\xi_i\}$ are i.i.d. copies of ξ.

Let $\bar{Y}_n := \sum_{k=0}^{n} Z_k$, i.e. the total population size up to generation n. Then, it is easy to get

$$P(\bar{Y}_n[n] \geq \lambda n^2) \leq c \exp(-c'\lambda), \quad P(\bar{Y}_n[n] \leq \lambda n^2) \leq c \exp(-c'/\sqrt{\lambda}).$$

We can obtain similar estimates for $\tilde{Y}_n[n]$ so (i) holds.

(ii) For $x, y \in \mathcal{G}$, let $\gamma(x, y)$ be the unique geodesic between x and y. Let $D(x)$ be the descendants of x and $D_r(x) = \{y \in D(x) : d(x, y) = r\}$. Let H be the backbone and $b = b_{r/4} \in H$ be the point where $d(0, b) = r/4$. Define

$$A := \cup_{z \in \gamma(0, b) \setminus \{b\}} (D_{r/4}(z) \setminus H), \quad A^* = \{z \in A : D_{r/4}(z) \neq \emptyset\}.$$

Then, any path from 0 to $B(0, r)^c$ must cross $A^* \cup \{b\}$, so that $M(0, r) \leq |A^*| + 1$. Define $p_r := P(z \in A^* | z \in A) = P(Z_{r/4} > 0) \leq c/r$. Let $\{\kappa_i\}$ be i.i.d. with distribution $\mathrm{Ber}(p_r)$ that are independent of $|A|$. Then we see that

$$|A^*| \overset{(d)}{=} \sum_{i=1}^{|A|} \kappa_i, \quad |A| \overset{(d)}{\leq} \tilde{Z}_{r/4}[r/4].$$

Using these, it is easy to obtain $P(|A^*| > m) \leq e^{-cm}$, so (ii) holds.   □

We next give the key lemma for the proof of Theorem 7.1.2 (ii)

Lemma 7.2.2. *For any $\varepsilon \in (0, 1)$,*

$$\limsup_{n \to \infty} \frac{V(0, n)}{n^2 (\log \log n)^{1-\varepsilon}} = \infty, \quad \mathbb{P} - a.s.$$

Proof (Sketch). Let $D(x; z) = \{y \in D(x) : \gamma(x, y) \cap \gamma(x, z) = \{x\}\}$, and define

$$Z_n = |\{x : x \in D(y_i; y_{i+1}), d(x, y_i) \leq 2^{n-2}, 2^{n-1} \leq i \leq 2^{n-1} + 2^{n-2}\}|.$$

Thus Z_n is the number of descendants off the backbone, to level 2^{n-2}, of points y on the backbone between levels 2^{n-1} and $2^{n-1} + 2^{n-2}$. So $\{Z_n\}_n$ are independent, $|B(0, 2^n)| \geq Z_n$, and $Z_n \overset{(d)}{=} \tilde{Y}_{2^{n-2}}[2^{n-2}]$. It can be shown that $\tilde{Y}_n[n] \overset{(d)}{\geq} c_1 n^2 \mathrm{Bin}(n, p_1/n)$ for some $p_1 > 0$ so we have, if $a_n = (\log n)^{1-\varepsilon}$,

$$\mathbb{P}(|B(0, 2^n)| \geq a_n 4^n) \geq \mathbb{P}(Z_n \geq a_n 4^n) \geq P(\tilde{Y}_{2^{n-2}}[2^{n-2}] \geq a_n 4^n)$$
$$\geq P(\mathrm{Bin}(2^{n-2}, p_1 2^{-n+2}) \geq c_2 a_n) \geq c_3 e^{-c_2 a_n \log a_n} \geq c_3/n.$$

As Z_n are independent, the desired estimate follows by the second Borel-Cantelli Lemma.   □

Proof of Theorem 7.1.2 (ii). Let $a_n = V(0, 2^n) 2^{-2n}, t_n = 2^n V(0, 2^n) = a_n 2^{3n}$. Using (4.22) with $\alpha = 1$, we have $f_{rV(0,r)}(0) \leq c/V(0, r)$, so $p_{t_n}^\omega(0, 0) \leq c/V(0, 2^n) = ct_n^{-2/3}/a_n^{1/3}$. By Lemma 7.2.2, $a_n \geq (\log n)^{1-\varepsilon} \asymp (\log \log t_n)^{1-\varepsilon}$. We thus obtain the result.   □

7.3   Random Walk on IIC for Critical Oriented Percolation in $\mathbb{Z}^d$ $(d > 6)$

We first introduce the spread-out oriented percolation model. Let $d > 4$ and $L \geq 1$. We consider an oriented graph with vertices $\mathbb{Z}^d \times \mathbb{Z}_+$ and oriented bonds $\{((x, n), (y, n+1)) : n \geq 0, x, y \in \mathbb{Z}^d \text{ with } 0 \leq \|x - y\|_\infty \leq L\}$. We consider bond percolation on the graph. We write $(x, n) \to (y, m)$ if there is a sequence of open oriented bonds that connects (x, n) and (y, m). Let

$$C(x, n) = \{(y, m) : (x, n) \to (y, m)\}$$

and define $\theta(p) = \mathbb{P}_p(|C(0, 0)| = \infty)$. Then, there exists $p_c = p_c(d, L) \in (0, 1)$ such that $\theta(p) > 0$ for $p > p_c$ and $\theta(p) = 0$ for $p \le p_c$. In particular, there is no infinite cluster when $p = p_c$ (see [122, p. 369], [123]).

For this example, the construction of incipient infinite cluster is given by van der Hofstad et al. [137] for $d > 4$. Let $\tau_n(x) = \mathbb{P}_{p_c}((0, 0) \to (x, n))$ and $\tau_n = \sum_x \tau_n(x)$.

Proposition 7.3.1 ([137]).

(i) *There exists $L_0(d)$ such that if $L \ge L_0(d)$, then*

$$\exists \lim_{n \to \infty} \mathbb{P}_{p_c}(E | (0, 0) \to n) =: \mathbb{Q}_{\mathrm{IIC}}(E) \qquad \text{for any cylindrical events } E,$$

where $(0, 0) \to n$ means $(0, 0) \to (x, n)$ for some $x \in \mathbb{Z}^d$. Further, $\mathbb{Q}_{\mathrm{IIC}}$ can be extended uniquely to a probability measure on the Borel σ-field and $C(0, 0)$ is $\mathbb{Q}_{\mathrm{IIC}}$-a.s. an infinite cluster.

(ii) *There exists $L_0(d)$ such that if $L \ge L_0(d)$, then*

$$\exists \lim_{n \to \infty} \frac{1}{\tau_n} \sum_x \mathbb{P}_{p_c}(E \cap \{(0, 0) \to (x, n)\}) =: \mathbb{P}_{\mathrm{IIC}}(E), \ \forall \ \text{cylindrical events } E.$$

Further, $\mathbb{P}_{\mathrm{IIC}}$ can be extended uniquely to a probability measure on the Borel σ-field and $\mathbb{P}_{\mathrm{IIC}} = \mathbb{Q}_{\mathrm{IIC}}$.

Let us consider simple random walk on the IIC. It is proved in [32] that (5.1) in Theorem 5.1.1 holds for $\mathbb{P}_{\mathrm{IIC}}$ with $p(\lambda) = \lambda^{-1}$, $D = 2$ and $\alpha = 1$, so we have (5.2).

Note that although many of the arguments in Chap. 6 can be generalized to this oriented model straightforwardly, some are not, due to the orientedness. For example, it is not clear how to adapt Proposition 6.1.3 ii) to the oriented model. (The definition of $\Gamma(r)$ needs to be modified in order to take orientedness into account.) For the reference, we will briefly explain how the proof of (5.1) goes in [32], and explain why $d > 6$ is needed.

Volume Estimates. Let $Z_R := cV(0, R)/R^2$ and $Z = \int_0^1 dt \, W_t(\mathbb{R}^d)$, where W_t is the canonical measure of super-Brownian motion conditioned to survive for all time.

Proposition 7.3.2. *The following holds for $d > 4$ and $L \ge L_0(d)$.*

(i) $\lim_{R \to \infty} \mathbb{E}_{\mathrm{IIC}} Z_R^l = \mathbb{E} Z^l \le 2^{-l}(l + 1)!$ *for all $l \in \mathbb{N}$. In particular, $c_1 R^2 \le \mathbb{E}_{\mathrm{IIC}} V(0, R) \le c_2 R^2$ for all $R \ge 1$.*

(ii) $\mathbb{P}_{\mathrm{IIC}}(V(0, R)/R^2 < \lambda) \le c_1 \exp\{-c_2/\sqrt{\lambda}\}$ *for all $R, \lambda \ge 1$.*

Proof (Sketch).

(i) First, note that for $l \geq 1$ and R large,

$$\mathbb{E}_{\text{IIC}} Z_R^l \sim \mathbb{E}_{\text{IIC}}[(c' R^{-2}|B(0, R)|)^l] = (c' R^{-2})^l \mathbb{E}_{\text{IIC}}[(\sum_{n=0}^{R-1} \sum_{y \in \mathbb{Z}^d} I_{\{(0,0) \to (y,n)\}})^l].$$

(7.5)

For $l \geq 1$ and $\mathbf{m} = (m_1, \cdots, m_l)$, define the IIC $(l+1)$-points function as

$$\hat{\rho}_{\mathbf{m}}^{(l+1)} = \sum_{y_1, \cdots, y_l \in \mathbb{Z}^d} \mathbb{P}_{\text{IIC}}((0,0) \to (y_i, m_i), \forall i = 1, \cdots, l).$$

In [137], it is proved that for $\mathbf{t} = (t_1, \cdots, t_l) \in (0, 1]^l$,

$$\lim_{s \to \infty} (\frac{c'}{s})^l \hat{\rho}_{st}^{(l+1)} = \hat{M}_{1,\mathbf{t}}^{(l+1)} := \mathbb{N}(X_1(\mathbb{R}^d), X_{t_1}(\mathbb{R}^d), \cdots, X_{t_l}(\mathbb{R}^d)),$$

where $\hat{M}_{1,\mathbf{t}}^{(l+1)}$ is the $(l+1)$-st moment of the canonical measure $\mathbb{N}$ of super-BM X_t. So taking $R \to \infty$

$$\text{(RHS of (7.5))} = \frac{c'^l}{R^{2l}} \sum_{n_1=0}^{R-1} \cdots \sum_{n_l=0}^{R-1} \hat{\rho}_{n_1, \cdots, n_l}^{(l+1)} = \frac{1}{R} \sum_{n_1=0}^{R-1} \cdots \frac{1}{R} \sum_{n_l=0}^{R-1} \frac{c'^l}{R^l} \hat{\rho}_{Rt}^{(l+1)}$$

$$\to \int_0^1 dt_1 \cdots \int_0^1 dt_l \hat{M}_{1,\mathbf{t}}^{(l+1)} = \mathbb{E} Z^l$$

where $\mathbf{t} = (n_1 R^{-1}, \cdots n_l R^{-1})$.

(ii) By (i), we see that for any $\epsilon > 0$, there exists $\delta > 0$ such that $\mathbb{P}_{\text{IIC}}(V(0, R)/R^2 < \delta) < \epsilon$. Now the chaining argument gives the desired estimate. $\square$

From Proposition 7.3.2, we can verify the volume estimates in (5.1). As we see, we can obtain it for all $d > 4$.

Resistance Estimates. Recall that J_n is the set of lanes for R at level n $(n \leq R)$.

Proposition 7.3.3. *For $d > 6$, there exists $L_1(d) \geq L_0(d)$ such that for $L \geq L_1(d)$,*

$$\mathbb{E}_{\text{IIC}}(|J_n|) \leq c_1(a, d), \quad 0 < \forall n < aR, \quad 0 < \forall a < 1, \forall R \geq 1.$$

Note that similarly to (6.15), we have $R_{\text{eff}}(0, \partial B(0, R)) \geq \sum_{n=1}^R \frac{1}{|J_n|}$. So, once Proposition 7.3.3 is proved, we can prove $\mathbb{P}_{\text{IIC}}(R_{\text{eff}}(0, \partial B(R)) \leq \varepsilon R) \leq c\varepsilon$. Thus we can verify the resistance estimates in (5.1) and the proof is complete.

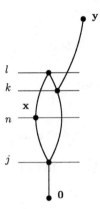

Fig. 7.1 Zigzag path

Proof of Proposition 7.3.3 is quite involved and use $d > 6$. Let us indicate why $d > 6$ is needed.

$$\mathbb{E}_{\mathrm{IIC}}|J_n| = \sum_{w,x \in \mathbb{Z}^d} \mathbb{P}_{\mathrm{IIC}}\left[(\mathbf{w}, \mathbf{x}) \in J_n\right]$$

$$= \frac{1}{A} \lim_{N \to \infty} \sum_{w,x,y \in \mathbb{Z}^d} \mathbb{P}_{p_c}\left[(\mathbf{w}, \mathbf{x}) \in J_n, \mathbf{0} \to \mathbf{y}\right],$$

where $\mathbf{w} = (w, n-1), \mathbf{x} = (x, n), \mathbf{y} = (y, N)$ and $A = \lim_{N \to \infty} \tau_N$. Let us consider one configuration in the event $\{(\mathbf{w}, \mathbf{x}) \in J_n, \mathbf{0} \to \mathbf{y}\}$ which is indicated by Fig. 7.1.

By [139], we have $\sup_{x \in \mathbb{Z}^d} \tau_n(x) \le K\beta(n+1)^{-d/2}$ for $n \ge 1$, and $\tau_n = A(1 + O(n^{(4-d)/2}))$ as $n \to \infty$. Using these and the BK inequality, the configuration in the above figure can be bounded above by

$$c \sum_{l=n}^{\infty} \sum_{k=n}^{l} \sum_{j=0}^{n} (l - j + 1)^{-d/2} \le c \sum_{l=n}^{\infty} \sum_{k=n}^{l} (l - n + 1)^{(2-d)/2}$$

$$\le c \sum_{l=n}^{\infty} (l - n + 1)^{(4-d)/2}$$

$$= c \sum_{m=1}^{\infty} m^{(4-d)/2} < \infty \quad \text{for } d > 6.$$

In order to prove Proposition 7.3.3, one needs to estimate more complicated zigzag paths efficiently.

(Open problem) The critical dimension for oriented percolation is 4. How does the random walk on IIC for oriented percolation behaves when $d = 5, 6$? One can ask the same question for the critical branching random walk. Quite recently, it is proved in [141] that the effective resistance between the origin and generation n of the incipient infinite oriented (and non-oriented) branching random walk in dimensions $d < 6$ is $O(n^{1-\alpha})$ for some $\alpha > 0$. So the critical dimension for the random walk on this model is 6.

7.4   Below Critical Dimension

We have seen various IIC models where simple random walk on the IIC enjoys the estimate (5.2) with $D = 2$ and $\alpha = 1$. This is a typical mean field behavior that may hold for high dimensions (above the critical dimension). It is natural to ask for the behavior of simple random walk on low dimensions. Very few is known. In this section, we will list up rigorous results in random media that are related in the sense that the behavior of the random walk/Brownian motion is anomalous, but not like high dimensional IIC. (The sub-linear behavior of the effective resistance in [141] for the critical branching random walk in dimensions $d < 6$ falls into this category, but since we already discussed it, we will not put it here.)

(i) Random walk on the IIC for two-dimensional critical percolation [149] and on the two-dimensional invasion percolation cluster [88]

In [149], Kesten shows the existence of IIC for two-dimensional critical percolation cluster. Further, he proves subdiffusive behavior of simple random walk on IIC in the following sense. Let $\{Y_n\}_n$ be a simple random walk on the IIC, then there exists $\epsilon > 0$ such that the $\mathbb{P}$-distribution of $n^{-\frac{1}{2}+\epsilon}d(0, Y_n)$ is tight. In [88], a quenched version of Kesten's result is established both for the IIC and the invasion percolation cluster. Let τ_n be the first exit time from the box of size $2n$ centered at the origin. Then there exists $\varepsilon > 0$ such that $\tau_n \geq n^{2+\varepsilon}$ a.s. both for the randomness of the media and realization of the random walk started at the origin. Some quantitative lower bound of ε is also obtained.

(ii) Random walk on the two-dimensional uniform spanning tree [34]

It is proved that (5.2) holds with $D = 8/5$ and $\alpha = 1$. Especially, it is shown that the spectral dimension of the random walk is $16/13$.

(iii) Brownian motion on the critical percolation cluster for the diamond lattice [131]

Brownian motion is constructed on the critical percolation cluster for the diamond lattice. Further, it is proved that the heat kernel enjoys continuous version of (5.2) with $\alpha = 1$ and some non-trivial D that is determined by the maximum eigenvalue of the matrix for the corresponding multi-dimensional branching process.

(iv) Random walk on non-intersecting two-sided random walk trace [201]

Let $\{\bar{S}_i\}_{i=1,2}$ be random walks on $\mathbb{Z}^d$ with $\bar{S}_i(0) = 0$ conditioned that $\bar{S}_1[0, \infty) \cap \bar{S}_2[1, \infty) = \emptyset$. In [201], it is proved that the simple random walk on $\bar{S}_1[0, \infty) \cup \bar{S}_2[1, \infty)$ is subdiffusive in the quenched level for $d = 2, 3$. One of the key ingredients of the proof is to estimate the number of global cut times up to time n.

(v) Random walk on the uniform infinite planar triangulation/quadrangulation [46, 128]

We first define uniform infinite planar triangulation/quadrangulation (UIPT and UIPQ). A planar map is a proper embedding (up to orientation-preserving homeomorphisms) of a finite connected graph in the two-dimensional sphere. We consider a rooted planar map, i.e. a distinguished oriented edge (called the root) is given. A triangulation (resp. quadrangulation) is a planar map whose faces all have degree 3 (resp. 4). For a rooted map (m, ρ) where ρ is the root, let $B_m(\rho, r)$ be the map formed by the faces of m that have at least one vertex at distance $< r$ from ρ. Given two rooted maps (m, ρ) and (m', ρ'), define their local distance to be $(1 + A)^{-1}$ where $A = \sup\{r \geq 1 : B_m(\rho, r) = B_{m'}(\rho', r')\}$. Let Q_n be a random rooted triangulation/quadrangulation uniformly distributed over the rooted quadrangulations with n faces. It is proved that Q_n converges in distribution (for the local distance) to a random infinite rooted triangulation/quadrangulation Q_∞, which is called the UIPT/UIPQ (see [14, 50, 164]). We note that random planar maps are very widely studied in probability, combinatorics and statistical physics; see [169] and the references therein.

In [46], it is proved that simple random walk on UIPQ is subdiffusive almost surely. Indeed, they prove that there exists $\kappa > 0$ such that

$$d(\rho, X_n) \leq n^{1/3}(\log n)^\kappa, \qquad \mathbb{P}\text{-a.s. for large } n,$$

where $\{X_n\}_n$ is the simple random walk started at $\rho \in Q_\infty$, and $d(\cdot, \cdot)$ is the graph distance. Very recently, it is proved in [128] that any distributional limit of finite planar graphs in which the degree of the root has an exponential tail is almost surely recurrent. In particular, UIPT/UIPQ are recurrent. However, sharp heat kernel estimates are not established in these models. In [12], it is shown that a ball of radius r has volume $r^{4+o(1)}$, so $d_f = 4$; the result of [46] shows $d_w \geq 3$, and it is conjectured that $d_w = 4$.

Back to the percolation model, it is believed that the critical dimension for percolation is 6. It would be very interesting to know the spectral dimension of simple random walk on IIC for the critical percolation cluster for $d < 6$. (Note that in this case even the existence of IIC is not proved except for $d = 2$.) The following numerical simulations (which we borrow from [44]) suggest that the Alexander–Orbach conjecture does not hold for $d \leq 5$.

$$d = 5 \Rightarrow d_s = 1.34 \pm 0.02, \quad d = 4 \Rightarrow d_s = 1.30 \pm 0.04,$$
$$d = 3 \Rightarrow d_s = 1.32 \pm 0.01, \quad d = 2 \Rightarrow d_s = 1.318 \pm 0.001.$$

7.5 Random Walk on Random Walk Traces and on Erdős-Rényi Random Graphs

In this section, we discuss the behavior of random walk in two random media, namely on random walk traces and on Erdős-Rényi random graphs. They are not IIC, but the technique discussed in Chap. 4 and Sect. 5.1 can be applied to some extent. We give a brief overview of the results.

Random Walk on Random Walk Traces. Let $X(\omega)$ be the trace of simple random walk on $\mathbb{Z}^d$, $d \geq 3$, started at 0. Let $\{Y_t^\omega\}_{t \geq 0}$ be the simple random walk on $X(\omega)$ and $p_n^\omega(\cdot, \cdot)$ be its heat kernel. It is known in general that if a Markov chain corresponding to a weighted graph is transient, then the simple random walk on the trace of the Markov chain is recurrent $\mathbb{P}$-a.s. (see [48]). The question is to have more detailed properties of the random walk when the initial graph is $\mathbb{Z}^d$. The following results show that it behaves like one-dimensional simple random walk when $d \geq 5$.

Theorem 7.5.1 ([82]). *Let $d \geq 5$, and let $\{B_t\}_{t \geq 0}, \{W_t^{(d)}\}_{t \geq 0}$ be independent standard Brownian motions on $\mathbb{R}$ and $\mathbb{R}^d$ respectively, both started at 0.*

(i) There exist $c_1, c_2 > 0$ such that

$$c_1 n^{-1/2} \leq p_{2n}^\omega(0,0) \leq c_2 n^{-1/2} \quad \text{for large } n, \quad \mathbb{P}\text{-a.e. } \omega.$$

(ii) There exists $\sigma_1 = \sigma_1(d) > 0$ such that $\{n^{-1/2} d^\omega(0, Y_{[tn]}^\omega)\}_{t \geq 0}$ converges weakly to $\{|B_{\sigma_1 t}|\}_{t \geq 0}$ $\mathbb{P}$-a.e. ω, where $d^\omega(\cdot, \cdot)$ is the graph distance on $X(\omega)$. Also, there exists $\sigma_2 = \sigma_2(d) > 0$ such that $\{n^{-1/4} Y_{[tn]}^\omega\}_{t \geq 0}$ converges weakly to $\{W_{|B_{\sigma_2 t}|}^{(d)}\}_{t \geq 0}$ $\mathbb{P}$-a.e. ω.

On the other hand, the behavior is different for $d = 3, 4$.

Theorem 7.5.2 ([199, 200]).

(i) Let $d = 4$. Then there exist $c_1, c_2 > 0$ and a slowly varying function ψ such that

$$c_1 n^{-\frac{1}{2}} (\psi(n))^{\frac{1}{2}} \leq p_{2n}^\omega(0,0) \leq c_2 n^{-\frac{1}{2}} (\psi(n))^{\frac{1}{2}} \quad \text{for large } n, \quad \mathbb{P}\text{-a.e. } \omega, \qquad (7.6)$$

$$n^{\frac{1}{4}} (\log n)^{\frac{1}{24} - \delta} \leq \max_{1 \leq k \leq n} |Y_k^\omega| \leq n^{\frac{1}{4}} (\log n)^{\frac{13}{12} + \delta} \quad \text{for large } n, \quad P_\omega^0\text{-a.s., } \mathbb{P}\text{-a.e. } \omega,$$

for any $\delta > 0$. Further, $\psi(n) \approx (\log n)^{-\frac{1}{2}}$, that is

$$\lim_{n \to \infty} \frac{\log \psi(n)}{\log \log n} = -\frac{1}{2}.$$

(ii) Let $d = 3$. Then there exists $\alpha > 0$ such that

$$p_{2n}^{\omega}(0,0) \leq n^{-\frac{10}{19}} (\log n)^{\alpha} \quad \text{for large } n, \quad \mathbb{P}\text{-a.s. } \omega.$$

These estimates suggest that the "critical dimension" for random walk on random walk trace for $\mathbb{Z}^d$ is 4. Note that some annealed estimates for the heat kernel (which are weaker than the above) are obtained in [82] for $d = 4$.

One of the key estimates to establish (7.6) is to obtain a sharp estimate for $\mathbb{E}[R_{\text{eff}}(0, S_n)]$ where $\{S_n\}_n$ is the simple random walk on $\mathbb{Z}^4$ started at 0. In [67], Burdzy and Lawler obtained

$$c_1(\log n)^{-\frac{1}{2}} \leq \frac{1}{n}\mathbb{E}[R_{\text{eff}}(0, S_n)] \leq c_2(\log n)^{-\frac{1}{3}} \qquad \text{for } d = 4. \qquad (7.7)$$

This comes from naive estimates $\mathbb{E}[L_n] \leq \mathbb{E}[R_{\text{eff}}(0, S_n)] \leq \mathbb{E}[A_n]$, where L_n is the number of cut points for $\{S_0, \cdots, S_n\}$, and A_n is the number of points for loop-erased random walk for $\{S_0, \cdots, S_n\}$. (The logarithmic estimates in (7.7) are those of $\mathbb{E}[L_n]$ and $\mathbb{E}[A_n]$.) In [200], Shiraishi proves the following.

$$\frac{1}{n}\mathbb{E}[R_{\text{eff}}(0, S_n)] \approx (\log n)^{-\frac{1}{2}} \qquad \text{for } d = 4,$$

which means the exponent that comes from the number of cut points is the right one. Intuitively the reason is as follows. Let $\{T_j\}$ be the sequence of cut times up to time n. Then the random walk trace near S_{T_j} and $S_{T_{j+1}}$ intersects typically when $T_{j+1} - T_j$ is large, i.e. there exists a "long range intersection". So $\mathbb{E}[R_{\text{eff}}(S_{T_j}, S_{T_{j+1}})] \approx 1$, and $\mathbb{E}[R_{\text{eff}}(0, S_n)] \asymp \mathbb{E}[\sum_{j=1}^{a_n} R_{\text{eff}}(S_{T_j}, S_{T_{j+1}})] \approx \mathbb{E}[a_n] \asymp n(\log n)^{-1/2}$, where $a_n := \sup\{j : T_j \leq n\}$.

Random Walk on Erdős-Rényi Random Graphs. Let $V_n := \{1, 2, \cdots, n\}$ be labeled vertices. Each bond $\{i, j\}$ $(i, j \in V_n)$ is open with probability $p \in (0, 1)$, independently of all the others. The realization $G(n, p)$ is the Erdős-Rényi random graph (cf. [98]). It is well-known (see [63]) that this model has a phase transition at $p = 1/n$ in the following sense. Let $\mathcal{C}_i^n$ be the i-th largest connected component. If $p \sim c/n$ with $c < 1$ then $|\mathcal{C}_1^n| = O(\log n)$, with $c > 1$ then $|\mathcal{C}_1^n| \asymp n$ and $|\mathcal{C}_j^n| = O(\log n)$ for $j \geq 2$, and if $p \sim c/n$ with $c = 1$ then $|\mathcal{C}_j^n| \asymp n^{2/3}$ for $j \geq 1$.

Now consider the finer scaling $p = 1/n + \lambda n^{-4/3}$ for fixed $\lambda \in \mathbb{R}$—the so-called critical window. Let $|\mathcal{C}_1^n|$ and S_1^n be the size and the surplus (i.e., the minimum number of edges which would need to be removed in order to obtain a tree) of $\mathcal{C}_1^n$. Then the well-known result by Aldous [6] says that $(n^{-2/3}|\mathcal{C}_1^n|, S_1^n)$ converges weakly to some random variables which are determined by a length of the largest excursion of reflected Brownian motion with drift and by some Poisson point process. Recently, Addario-Berry et al. [2] prove further that there exists a (random) compact metric space $\mathcal{M}_1$ such that $n^{-1/3}\mathcal{C}_1^n$ converges weakly to $\mathcal{M}_1$ in the Gromov-Hausdorff sense. (In fact, these results hold not only for $\mathcal{C}_1^n, S_1^n$ but also for the sequences $(C_1^n, C_2^n, \cdots)$, $(S_1^n, S_2^n, \cdots)$.) Here $\mathcal{M}_1$ can be constructed from a random real tree (given by the excursion mentioned above) by gluing a (random) finite number of points (chosen according to the Poisson point process)—see [2] for details.

Now we consider simple random walk $\{Y_m^{C_1^n}\}_m$ on C_1^n. The following results on the scaling limit and the heat kernel estimates are obtained by Croydon [83].

Theorem 7.5.3 ([83]). *There exists a diffusion process ("Brownian motion")* $\{B_t^{\mathcal{M}_1}\}_{t \geq 0}$ *on* $\mathcal{M}_1$ *such that* $\{n^{-1/3} Y_{[nt]}^{C_1^n}\}_{t \geq 0}$ *converges weakly to* $\{B_t^{\mathcal{M}_1}\}_{t \geq 0}$ $\mathbb{P}$-*distribution. Further, there exists a jointly continuous heat kernel* $p_t^{\mathcal{M}_1}(\cdot, \cdot)$ *for* $\{B_t^{\mathcal{M}_1}\}_{t \geq 0}$ *which enjoys the following estimates.*

$$c_1 t^{-2/3} (\ln_1 t^{-1})^{-\alpha_1} \exp\left(- c_2 \left(\frac{d(x,y)^3}{t}\right)^{1/2} \left(\ln_1\left(\frac{d(x,y)}{t}\right)\right)^{\alpha_2}\right) \leq p_t^{\mathcal{M}_1}(x,y)$$

$$\leq c_3 t^{-2/3} (\ln_1 t^{-1})^{1/3} \exp\left(- c_4 \left(\frac{d(x,y)^3}{t}\right)^{1/2} \left(\ln_1\left(\frac{d(x,y)}{t}\right)\right)^{-\alpha_3}\right),$$

for all $x, y \in \mathcal{M}_1, t \leq 1$, *where* $\ln_1 x := 1 \vee \log x$, $c_1, \cdots, c_4$ *are positive random constants and* $\alpha_1, \alpha_2, \alpha_3$ *are positive non-random constants.*

The same results hold for C_i^n and $\mathcal{M}_i$ for each $i \in \mathbb{N}$ (with constants depending on i).

7.6 Mixing Times and Cover Times

In this section, we introduce mixing times and cover times for finite graphs, and explain some recent results on the behavior of them for some sequences of finite (random) graphs.

Mixing Times. Let (X, μ) be a finite weighted graph. Let $\pi(\cdot) = \mu(\cdot)/\mu(X)$ be the invariant probability measure and let $p_n(\cdot, \cdot)$ be the heat kernel as defined in (2.3). For $p \in [1, \infty]$, the L^p-mixing time of X and L^p-mixing time of X at $\rho \in X$ is defined by

$$t_{\text{mix}}^p(X) := \inf\left\{m > 0 : \sup_{x \in X} D_p(x, m) \leq 1/4\right\},$$

$$t_{\text{mix}}^p(\rho) := \inf\{m > 0 : D_p(\rho, m) \leq 1/4\},$$

where $D_p(\rho, m) := \left\|\frac{p_m(\rho, y) + p_{m+1}(\rho, y)}{2} - 1\right\|_{L^p(\pi)}$. Note that when $p = 1$, $t_{\text{mix}}^1(X)$ coincides with the mixing time defined as

$$T_{\text{mix}}(X) = \min\{n : \|P^x(Y_n \in \cdot) - \pi(\cdot)\|_{\text{TV}} \leq 1/8, \ \forall x \in X\},$$

where $\|\mu - \nu\|_{\mathrm{TV}} = \max_{B \subset X} |\mu(B) - \nu(B)|$. Mixing times on non-random and random graphs have been extensively studied (see [170, 172, 181] for the history and recent developments). For percolation on finite graphs, sharp estimates of mixing times for random walks inside the so-called scaling window is established in [186], and using their result, [135, Corollary 1.3] gives more detailed estimates for critical percolation on the high-dimensional torus (both in the sense of "in probability for $\mathbb{P}$"). In [86], a notion of spectral Gromov-Hausdorff convergence is introduced for a sequence of finite graphs, and it is proved that under the spectral Gromov-Hausdorff convergence, suitably rescaled L^p-mixing times of the corresponding Markov chains converge to that of the diffusion process on the Gromov-Hausdorff limit of the (suitably rescaled) graphs. In particular, for the Erdős-Rényi random graphs in the critical window, fixing ρ^N in the N-level random graph, it is proved that

$$N^{-1} t_{\mathrm{mix}}^p (\rho^N) \to t_{\mathrm{mix}}^p (\rho), \qquad \text{in distribution},$$

where $t_{\mathrm{mix}}^p (\rho) \in (0, \infty)$ is the L^p-mixing time of the Brownian motion on $\mathcal{M}_1$ started from ρ, discussed in Theorem 7.5.3.

Cover Times and Maximum of the Gaussian Free Field. Let (X, μ) be a finite weighted graph. For simplicity, let us assume that each bond has conductance 1. The cover time for the corresponding Markov chain is defined by

$$t_{cov}(X) := \max_{x \in X} E^x [\max_{y \in X} \sigma_y(X)], \tag{7.8}$$

where $\sigma_y(X)$ is the first hitting time to y for the corresponding Markov chain. Cover times of finite graphs have been extensively studied by computer scientists (see for example, [29] for references). In [29], some useful geometric sufficient condition for the upper bound of the cover time is given in terms of the resistance metric. Using this, the following is proved (see [29, Theorem 3.1]): let $|X| = n$ and consider the percolation on the graph. If Proposition 6.1.3 holds with respect to the percolation probability $\mathbb{P}_p$, then for any $L, \delta > 0$, there exists $M > 0$ such that the following holds.

$$\mathbb{P}_p(\exists \mathcal{C} \text{ with } |\mathcal{C}| \geq Ln^{2/3} \text{ and } t_{cov}(\mathcal{C}) \notin [M^{-1}n, Mn]) < \delta.$$

Proposition 6.1.3 can be verified for various critical percolation models (including high dimensional tori), so $t_{cov}(\mathcal{C})$ is of order n with high probability [29, Theorem 1.3]. In the paper, estimates of cover times for Erdős-Rényi random graphs near the critical window are also obtained.

Cover times are closely related to the maximum of the Gaussian free field. Let $B \subset X$ be such that $B \neq X$. The Gaussian free field on X with boundary B is the zero-mean Gaussian field $\{\eta_z\}_{z \in X}$ with covariance given by the Green density of the Markov chain killed on exiting B. Using (4.10), we have $E[(\eta_x - \eta_y)^2] = R_{\mathrm{eff}}^B(x, y)$, where $R_{\mathrm{eff}}^B(\cdot, \cdot)$ is defined as in (4.9).

Let B be one point (otherwise, we should shorten B to one point and consider the Markov chain on the shorting network). In [92, Theorem 1.3], the following is proved: there exists $c_1, c_2 > 0$ such that

$$c_1|E| \cdot (\mathbb{E} \max_{x \in X} \eta_x)^2 \leq t_{cov}(X) \leq c_2|E| \cdot (\mathbb{E} \max_{x \in X} \eta_x)^2, \tag{7.9}$$

where $|E|$ is the number of edges for the graph X.

Chapter 8
Random Conductance Model

In this chapter, we will discuss recent developments on the random conductance model (RCM). We will mainly discuss the quenched invariance principle, i.e. the functional central limit theorem which is almost sure with respect to the randomness of the environments.

We note that there is a very nice survey on the recent progress on the RCM by Biskup [55]. Random walk on RCM is a special case of random walk in random environment (RWRE), in the sense that the random walk is reversible. The subject of RWRE has a long history; we refer to [64, 94, 209, 213, 214] for overviews and recent developments of this field. Further, we refer to [143, 157] and the references therein for the history and methods of homogenization for diffusions in random environments.

8.1 Overview

We first recall the constant speed random walk (CSRW) and the variable speed random walk (VSRW) that are introduced in Remark 2.1.7. Let (X, μ) be a weighted graph. Both CSRW and VSRW are continuous time Markov chains with transition probability $P(x, y) = \mu_{xy}/\mu_x$. For CSRW, the holding time is the exponential distribution with mean 1 for each point, and for VSRW, the holding time at x is the exponential distribution with mean μ_x^{-1} for each $x \in X$. The corresponding discrete Laplace operators are given in (2.5), (2.11), which we rewrite here.

$$\mathcal{L}_C f(x) := \mathcal{L}f(x) = \frac{1}{\mu_x} \sum_y (f(y) - f(x))\mu_{xy},$$

$$\mathcal{L}_V f(x) = \sum_y (f(y) - f(x))\mu_{xy}.$$

T. Kumagai, *Random Walks on Disordered Media and their Scaling Limits*, Lecture Notes in Mathematics 2101, DOI 10.1007/978-3-319-03152-1_8,
© Springer International Publishing Switzerland 2014

Recall also that for each f, g that have finite support, we have

$$\mathcal{E}(f, g) = -(\mathcal{L}_V f, g)_\nu = -(\mathcal{L}_C f, g)_\mu,$$

where ν is a measure on X such that $\nu(A) = |A|$ for all $A \subset X$.

Now consider $\mathbb{Z}^d$, $d \geq 2$ and let E_d be the set of non-oriented nearest neighbor bonds and let the conductance $\{\mu_e : e \in E_d\}$ be stationary and ergodic on a probability space $(\Omega, \mathcal{F}, \mathbb{P})$. For each $\omega \in \Omega$, let $(\{Y_t\}_{t \geq 0}, \{P_\omega^x\}_{x \in \mathbb{Z}^d})$ be either the CSRW or VSRW and define

$$q_t^\omega(x, y) = P_\omega^x(Y_t = y)/\theta_y$$

be the heat kernel of $\{Y_t\}_{t \geq 0}$ where θ is either ν or μ. This model is called the random conductance model (RCM for short).

We are interested in the long time behavior of $\{Y_t\}_{t \geq 0}$, especially in obtaining the heat kernel estimates for $q_t^\omega(\cdot, \cdot)$ and a quenched invariance principle (to be precise, quenched functional central limit theorem (quenched FCLT for short)) for $\{Y_t\}_{t \geq 0}$. Note that when $\mathbb{E}\mu_e < \infty$, it was proved in the 1980s that $\varepsilon Y_{t/\varepsilon^2}$ converges as $\varepsilon \to 0$ to Brownian motion on $\mathbb{R}^d$ with covariance $\sigma^2 I$ in law under $\mathbb{P} \times P_\omega^0$ with the possibility $\sigma = 0$ for some cases (see [91, 154, 159]). This is sometimes referred as the annealed (averaged) invariance principle. Let us make these more precise. Let $T > 0$ and F be a bounded continuous function on $D([0, T], \mathbb{R}^d)$. For $\omega \in \Omega$, set $\Psi_\varepsilon := E_\omega^0 F(\varepsilon Y_{\cdot/\varepsilon^2})$ and $\Psi_0 := E_{BM} F(\sigma W_\cdot)$, where (W, P_{BM}) is a Brownian motion started at 0. Then the quenched invariance principle (quenched FCLT) means $\Psi_\varepsilon \to \Psi_0$ in $\mathbb{P}$-a.s. for all F, whereas the annealed (averaged) invariance principle (annealed (averaged) FCLT) means $\mathbb{E}[\Psi_\varepsilon] \to \Psi_0$ for all F. One can define another convergence which is an intermediate of the two; a weak FCLT states that $\Psi_\varepsilon \to \Psi_0$ in $\mathbb{P}$-probability for all F. When $\mathbb{E}\mu_e < \infty$, the weak FCLT was established in [91, 154] for general stationary ergodic environments. In general, such a convergence in probability cannot be implied by the convergence of the average. However, in [24] it is proved that under some mild condition, the weak FCLT and the annealed (averaged) FCLT are equivalent.

From now on, we will discuss the case when $\{\mu_e : e \in E_d\}$ are i.i.d. unless otherwise indicated. If $p_+ := \mathbb{P}(\mu_e > 0) < p_c(\mathbb{Z}^d)$ where $p_c(\mathbb{Z}^d)$ is the critical probability for bond percolation on $\mathbb{Z}^d$, then $\{Y_t\}_{t \geq 0}$ is confined to a finite set $\mathbb{P} \times P_\omega^x$-a.s., so we consider the case $p_+ > p_c(\mathbb{Z}^d)$. Under the condition, there exists unique infinite connected component of edges with strictly positive conductances, which we denote by $\mathcal{C}_\infty$. Typically, we will consider the case where $0 \in \mathcal{C}_\infty$, namely we consider $\mathbb{P}(\cdot | 0 \in \mathcal{C}_\infty)$. With some abuse of notation, we will use notation $\mathbb{P}$ both for (the original) $\mathbb{P}$ and for $\mathbb{P}(\cdot | 0 \in \mathcal{C}_\infty)$.

We will consider the following cases, and discuss quenched heat kernel estimates and the quenched invariance principle for the corresponding Markov chain on $\mathcal{C}_\infty$:

- Case 0: $c^{-1} \leq \mu_e \leq c$ for some $c \geq 1$ (uniform elliptic case),
- Case 1: $0 \leq \mu_e \leq c$ for some $c > 0$,

- Case 2: $c \leq \mu_e < \infty$ for some $c > 0$.
- Case 3: $0 \leq \mu_e < \infty$.

(Of course, Case 0 is the special case of Case 1 and Case 2, and Case 3 contains all the cases. Since there are different interesting aspects, we split Case 1 and 2 and discuss separately.)

Case 0 and the Case of Supercritical Percolation. For Case 0, the following two-sided quenched Gaussian heat kernel estimates

$$c_1 t^{-d/2} \exp(-c_2|x - y|^2/t) \leq q_t^\omega(x, y) \leq c_3 t^{-d/2} \exp(-c_4|x - y|^2/t) \quad (8.1)$$

hold $\mathbb{P}$-a.s. for $t \geq |x - y|$ by the result in [89], and the quenched invariance principle is proved in [202]. When $\mu_e \in \{0, 1\}$, which is a special case of Case 1, the corresponding Markov chain is a random walk on supercritical percolation clusters. In this case, isoperimetric inequalities are proved in [180] (see also [190]), two-sided quenched Gaussian long time heat kernel estimates are obtained in [18] (precisely, (8.1) holds for $1 \vee S_x(\omega) \vee |x - y| \leq t$ where $\{S_x\}_{x \in \mathbb{Z}^d}$ satisfies $\mathbb{P}(S_x \geq n, x \in \mathcal{C}(0)) \leq c_1 \exp(-c_2 n^{\varepsilon_d})$ for some $\varepsilon_d > 0$), and the quenched invariance principle is proved in [202] for $d \geq 4$ and later extended to all $d \geq 2$ in [51, 179].

Case 1 This case is treated in [52, 59, 104, 178] for $d \geq 2$. (Note that the papers [52, 59] consider a discrete time random walk and [104, 178] considers CSRW. In fact, one can see that there is not a big difference between CSRW and VSRW in this case, as we will see in Theorem 8.1.2.)

Heat Kernel Estimates. In [52, 104], it is proved that Gaussian heat kernel bounds do not hold in general and anomalous behavior of the heat kernel is established for d large (see also [65]). In [104], Fontes and Mathieu consider VSRW on $\mathbb{Z}^d$ with conductance given by $\mu_{xy} = \omega(x) \wedge \omega(y)$ where $\{\omega(x) : x \in \mathbb{Z}^d\}$ are i.i.d. with $\omega(x) \leq 1$ for all x and

$$\mathbb{P}(\omega(0) \leq s) \sim s^\gamma \quad \text{as } s \downarrow 0, \quad (8.2)$$

for some $\gamma > 0$. They prove the following anomalous annealed heat kernel behavior.

$$\lim_{t \to \infty} \frac{\log \mathbb{E}[P_\omega^0(Y_t = 0)]}{\log t} = -(\frac{d}{2} \wedge \gamma). \quad (8.3)$$

In [66], they consider the conductance model where (8.2) holds for μ_e (instead of $\omega(0)$). They show that for $\gamma > 1/4$,

$$P_\omega^0(Y_t = 0) \leq c t^{-d/2}, \quad \mathbb{P}\text{-a.s., for large } t,$$

both for the CSRW and the VSRW. If $\gamma < 1/4$, for VSRW one can easily see (see [66, Remark 1.2 (3)]) the following annealed estimate: $\mathbb{E}[P_\omega^0(Y_t = 0)] \geq c t^{-2d\gamma}$,

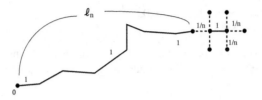

Fig. 8.1 Trap by small conductance

so $\gamma = 1/4$ is the critical value for the annealed return probability for the VSRW. Compare this with (8.3).

We now state the main results in [52] (together with the results in [56, 58] for $d = 4$). Here we consider discrete time Markov chain with transition probability $\{P(x, y) : x, y \in \mathbb{Z}^d\}$ and denote by $P_\omega^n(0, 0)$ the heat kernel for the Markov chain, which (in this case) coincides with the return probability for the Markov chain started at 0 to 0 at time n.

Theorem 8.1.1.   (i) *For $\mathbb{P}$-a.e. ω, there exists $C_1(\omega) < \infty$ such that for each $n \geq 1$,*

$$P_\omega^n(0, 0) \leq C_1(\omega) \begin{cases} n^{-d/2}, & d = 2, 3, \\ n^{-2} \log n, & d = 4, \\ n^{-2}, & d \geq 5. \end{cases} \tag{8.4}$$

Further, for $d \geq 5$, $\lim_{n \to \infty} n^2 P_\omega^n(0, 0) = 0$ $\mathbb{P}$-a.s., and for $d = 4$, $\lim_{n \to \infty} \frac{n^2}{\log n} P_n^\omega(0, 0) = 0$ $\mathbb{P}$-a.s.

(ii) *Let $d \geq 5$ and $\kappa > 1/d$. There exists an i.i.d. law $\mathbb{P}$ on bounded nearest-neighbor conductances with $p_+ > p_c(d)$ and $C_2(\omega) > 0$ such that for a.e. $\omega \in \{|\mathcal{C}(0)| = \infty\}$,*

$$P_\omega^{2n}(0, 0) \geq C_2(\omega) n^{-2} \exp(-(\log n)^\kappa), \quad \forall n \geq 1.$$

(iii) *Let $d \geq 4$. For any increasing sequence $\{\lambda_n\}_{n \in \mathbb{N}}$, $\lambda_n \to \infty$, there exists an i.i.d. law $\mathbb{P}$ on bounded nearest-neighbor conductances with $p_+ > p_c(d)$ and $C_3(\omega) > 0$ such that for a.e. $\omega \in \{|\mathcal{C}(0)| = \infty\}$,*

$$P_\omega^{2n}(0, 0) \geq C_3(\omega) n^{-2} \lambda_n^{-1} \quad \text{for } d \geq 5,$$
$$P_\omega^{2n}(0, 0) \geq C_3(\omega) n^{-2} (\log n) \lambda_n^{-1} \quad \text{for } d = 4.$$

along a subsequence that does not depend on ω.

As we can see, Theorem 8.1.1 (ii), (iii) shows anomalous behavior of the Markov chain for $d \geq 4$. We will give a key idea of the proof of (ii) here and give complete proof of it in Sect. 8.3.

Suppose we can show that for large n, there is a box of side length ℓ_n centered at the origin such that in the box a bond with conductance 1 ("strong" bond) is separated from other sites by bonds with conductance $1/n$ ("weak" bonds), and at least one of the "weak" bonds is connected to the origin by a path of bonds with conductance 1 within the box (see Fig. 8.1). Then the probability that the walk is back to the origin at time n is bounded below by the probability that the walk goes directly towards the above place (which costs $e^{O(\ell_n)}$ of probability) then crosses the weak bond (which costs $1/n$), spends time $n - 2\ell_n$ on the strong bond (which costs only $\Theta(1)$ of probability), then crosses a weak bond again (another $1/n$ term) and then goes back to the origin on time (another $e^{O(\ell_n)}$ term). The cost of this strategy is $\Theta(1)e^{O(\ell_n)}n^{-2}$ so if one can take $\ell_n = o(\log n)$ then we get the leading order n^{-2}.

Quenched Invariance Principle. For $t \geq 0$, let $\{Y_t\}_{t\geq 0}$ be either CSRW or VSRW and define

$$Y_t^{(\varepsilon)} := \varepsilon Y_{t/\varepsilon^2}. \tag{8.5}$$

In [59, 178], the following quenched invariance principle is proved.

Theorem 8.1.2. *(i) Let $\{Y_t\}_{t\geq 0}$ be the VSRW. Then $\mathbb{P}$-a.s. $Y^{(\varepsilon)}$ converges (under P_ω^0) in law to Brownian motion on $\mathbb{R}^d$ with covariance $\sigma_V^2 I$ where $\sigma_V > 0$ is non-random.*

(ii) Let $\{Y_t\}_{t\geq 0}$ be the CSRW. Then $\mathbb{P}$-a.s. $Y^{(\varepsilon)}$ converges (under P_ω^0) in law to Brownian motion on $\mathbb{R}^d$ with covariance $\sigma_C^2 I$ where $\sigma_C^2 = \sigma_V^2/(2d\mathbb{E}\mu_e)$.

Case 2 This case is treated in [28] for $d \geq 2$.

Heat Kernel Estimates. The following heat kernel estimates for the VSRW is proved in [28]. (We do not give a proof here.) Note that $\mathcal{C}_\infty = \mathbb{Z}^d$ in this case.

Theorem 8.1.3. *Let $q_t^\omega(x, y)$ be the heat kernel for the VSRW and let $\eta \in (0, 1)$. Then, there exist constants $c_1, \cdots, c_{11} > 0$ (depending on d and the distribution of μ_e) and a family of random variables $\{U_x\}_{x\in\mathcal{C}_\infty}$ with*

$$\mathbb{P}(U_x \geq n) \leq c_1 \exp(-c_2 n^\eta),$$

such that the following hold.

(a) For all $x, y \in \mathcal{C}_\infty$ and $t > 0$,

$$q_t^\omega(x, y) \leq c_3 t^{-d/2}.$$

(b) For $x, y \in \mathcal{C}_\infty$ and $t > 0$ with $|x - y| \vee t^{1/2} \geq U_x$,

$$q_t^\omega(x, y) \leq c_3 t^{-d/2} \exp(-c_4 |x - y|^2/t) \quad \text{if } t \geq |x - y|,$$
$$q_t^\omega(x, y) \leq c_3 \exp(-c_4 |x - y|(1 \vee \log(|x - y|/t))) \quad \text{if } t \leq |x - y|.$$

(c) For $x, y \in \mathcal{C}_\infty$ and $t > 0$ with $t \geq U_x^2 \vee |x - y|^{1+\eta}$,

$$q_t^\omega(x, y) \geq c_5 t^{-d/2} \exp(-c_6 |x - y|^2/t).$$

(d) For $x, y \in C_\infty$ and $t > 0$ with $t \geq c_7 \vee |x - y|^{1+\eta}$,

$$c_8 t^{-d/2} \exp(-c_9 |x - y|^2 / t) \leq \mathbb{E}[q_t^\omega(x, y)] \leq c_{10} t^{-d/2} \exp(-c_{11}|x - y|^2 / t).$$

Quenched Invariance Principle. For $t \geq 0$, define $Y_t^{(\varepsilon)}$ as in (8.5). Then the following quenched invariance principle is proved in [28].

Theorem 8.1.4. *(i) Let $\{Y_t\}_{t \geq 0}$ be the VSRW. Then $\mathbb{P}$-a.s. $Y^{(\varepsilon)}$ converges (under P_ω^0) in law to Brownian motion on $\mathbb{R}^d$ with covariance $\sigma_V^2 I$ where $\sigma_V > 0$ is non-random.*

(ii) Let $\{Y_t\}_{t \geq 0}$ be the CSRW. Then $\mathbb{P}$-a.s. $Y^{(\varepsilon)}$ converges (under P_ω^0) in law to Brownian motion on $\mathbb{R}^d$ with covariance $\sigma_C^2 I$ where $\sigma_C^2 = \sigma_V^2 / (2d\mathbb{E}\mu_e)$ if $\mathbb{E}\mu_e < \infty$ and $\sigma_C^2 = 0$ if $\mathbb{E}\mu_e = \infty$.

Local Central Limit Theorem. In [31], a sufficient condition is given for the quenched local limit theorem to hold (see [85] for a generalization to sub-Gaussian type local CLT). Using the results, the following local CLT is proved in [28]. (We do not give a proof here.)

Theorem 8.1.5. *Let $q_t^\omega(x, y)$ be the VSRW and write*

$$k_t(x) = (2\pi t \sigma_V^2)^{-d/2} \exp(-|x|^2 / (2\sigma_V^2 t))$$

where σ_V is as in Theorem 8.1.4 (i). Let $T > 0$, and for $x \in \mathbb{R}^d$, write $[x] = ([x_1], \cdots, [x_d])$. Then

$$\lim_{n \to \infty} \sup_{x \in \mathbb{R}^d} \sup_{t \geq T} |n^{d/2} q_{nt}^\omega(0, [n^{1/2}x]) - k_t(x)| = 0, \quad \mathbb{P}\text{-a.s.}$$

The key idea of the proof is as follows: one can prove the parabolic Harnack inequality using Theorem 8.1.3. This implies uniform Hölder continuity of $n^{d/2} q_{nt}^\omega(0, [n^{1/2} \cdot])$, which, together with Theorem 8.1.4 implies pointwise uniform convergence.

For the case of simple random walk on the supercritical percolation, this local CLT is proved in [31]. Note that in general when $\mu_e \leq c$, such a local CLT does NOT hold because of the anomalous behavior of the heat kernel and the quenched invariance principle.

Case 3 In a recent paper [10], they obtain the quenched invariance principle for this unified case for $d \geq 2$. The precise statement is exactly the same as that of Theorem 8.1.4. In the paper, they also obtain several quenched Green density estimates for $d \geq 3$. See Sect. 8.9 about how to obtain Theorem 8.1.4 for the unified case.

More About CSRW with $\mathbb{E}\mu_e = \infty$. According to Theorem 8.1.4 (ii), one does not have the usual central limit theorem for CSRW with $\mathbb{E}\mu_e = \infty$ in the sense the scaled process degenerates as $\varepsilon \to 0$. A natural question is what is the right scaling order and what is the scaling limit. The answers are given in [25, 36, 70] for the case of heavy-tailed environments. Let $\{\mu_e\}$ satisfy

$$\mathbb{P}(\mu_e \geq c_1) = 1, \quad \mathbb{P}(\mu_e \geq u) = c_2 u^{-\alpha}(1 + o(1)) \text{ as } u \to \infty, \quad (8.6)$$

for some constants $c_1, c_2 > 0$ and $\alpha \in (0, 1]$.

In order to state the result, we first introduce the Fractional-Kinetics (FK) process and the Fontes-Isopi-Newman (FIN) diffusion [103].

Definition 8.1.6. Let $\{B_d(t)\}$ be a standard d-dimensional Brownian motion started at 0.

(i) For $\alpha \in (0, 1)$, let $\{V_\alpha(t)\}_{t \geq 0}$ be an α-stable subordinator independent of $\{B_d(t)\}$, which is determined by $\mathbb{E}[\exp(-\lambda V_\alpha(t))] = \exp(-t\lambda^\alpha)$. Let $V_\alpha^{-1}(s) := \inf\{t : V_\alpha(t) > s\}$ be the rightcontinuous inverse of $V_\alpha(t)$. We define the fractional-kinetics process $\mathbf{FK}_{d,\alpha}$ by

$$\mathbf{FK}_{d,\alpha}(s) = B_d(V_\alpha^{-1}(s)), \quad s \in [0, \infty).$$

(ii) Let (x_i, v_i) on $\mathbb{R} \times \mathbb{R}_+$ be an inhomogeneous Poisson point process with intensity $dx\alpha v^{-1-\alpha}dv$ and let ρ be the random discrete measure defined by $\rho := \sum_i v_i \delta_{x_i}$. Set $\phi_\rho(t) := \int_{\mathbb{R}} \ell(t, y)\rho(dy)$ where $\ell(\cdot, \cdot)$ is the local time of the Brownian motion $\{B_1(t)\}$. We define the Fontes-Isopi-Newman (FIN) diffusion by

$$Z(s) = B_1(\phi_\rho^{-1}(s)), \quad s \in [0, \infty).$$

In other words, the FIN diffusion is a diffusion process (with $Z(0) = 0$) that can be expressed as a time change of Brownian motion with the speed measure ρ.

Note that the clock process V_α for the FK process is independent of Brownian motion, whereas the clock process ϕ_ρ for the FIN diffusion depends on Brownian motion.

The FK process is a non-Markovian process, which is γ-Hölder continuous for all $\gamma < \alpha/2$ and is self-similar, i.e. $\mathbf{FK}_{d,\alpha}(\cdot) \overset{(d)}{=} \lambda^{-\alpha/2}\mathbf{FK}_{d,\alpha}(\lambda\cdot)$ for all $\lambda > 0$. The density of the process $p(t, x)$ started at 0 satisfies the fractional-kinetics equation

$$\frac{\partial^\alpha}{\partial t^\alpha} p(t, x) = \frac{1}{2}\Delta p(t, x) + \delta_0(x)\frac{t^{-\alpha}}{\Gamma(1 - \alpha)}.$$

This process is well-known in physics literatures, see [212] for details.

Theorem 8.1.7. Let $\{Y_t\}_{t \geq 0}$ be the CSRW of RCM that satisfies (8.6).

(i) ([25]) Let $d \geq 3$, $\alpha \in (0, 1)$ in (8.6) and let $Y_t^{(\varepsilon)} := \varepsilon Y_{t/\varepsilon^2/\alpha}$. Then $\mathbb{P}$-a.s. $Y^{(\varepsilon)}$ converges (under P_ω^0) in law to a multiple of the fractional-kinetics process $c \cdot \mathbf{FK}_{d,\alpha}$ on $D([0, \infty), \mathbb{R}^d)$ equipped with the Skorokhod J_1-topology.

(ii) ([36]) *Let $d \geq 3$, $\alpha = 1$ in (8.6) with $c_1 = c_2 = 1$ and let $Y_t^{(\varepsilon)} := \varepsilon Y_{t \log(1/\varepsilon)/\varepsilon^2}$. Then $\mathbb{P}$-a.s. $Y^{(\varepsilon)}$ converges (under P_ω^0) in law to Brownian motion on $\mathbb{R}^d$ with covariance $\sigma_C^2 I$ where $\sigma_C = 2^{-1/2}\sigma_V > 0$.*

(iii) ([70]) *Let $d = 2$, $\alpha \in (0, 1)$ in (8.6) and let $Y_t^{(\varepsilon)} := \varepsilon Y_{t(\log(1/\varepsilon))^{1-1/\alpha}/\varepsilon^{2/\alpha}}$. Then the conclusion of (i) holds.*

(iv) ([70]) *Let $d = 1$, $\alpha \in (0, 1)$ in (8.6) and let $Y_t^{(\varepsilon)} := \varepsilon Y_{c_* c_\varepsilon t/\varepsilon}$, where $c_* = \mathbb{E}[\mu_e^{-1}]$ and*

$$c_\varepsilon := \inf\{t \geq 0 : \mathbb{P}(\mu_e > t) \leq \varepsilon\} = \varepsilon^{-1/\alpha}(1 + o(1)).$$

Then, $Y^{(\varepsilon)}$ converges in law to the FIN diffusion $Z(t)$ under $\mathbb{P} \times P_0^\mu$.

Remark 8.1.8. (i) In [41], a scaling limit theorem similar to Theorem 8.1.7 (i) is shown for symmetric Bouchaud's trap model (BTM) for $d \geq 2$. Let $\{\tau_x\}_{x \in \mathbb{Z}^d}$ be a positive i.i.d. and let $a \in [0, 1]$ be a parameter. Define a random weight (conductance) by

$$\mu_{xy} = \tau_x^a \tau_y^a \qquad \text{if } x \sim y,$$

and let $\mu_x = \tau_x$ be the measure. Then, the BTM is the CSRW with the transition probability $\mu_{xy}/\sum_y \mu_{xy}$ and the measure μ_x. If $a = 0$, then the BTM is a time change of the simple random walk on $\mathbb{Z}^d$ and it is called symmetric BTM (this is THE BTM), while non-symmetric refers to the general case $a \neq 0$. (This terminology is a bit confusing. Note that the Markov chain for the BTM is symmetric (reversible) with respect to μ for all $a \in [0, 1]$.) In [183], Theorem 8.1.7 (i) is shown for BTM for all $a \in [0, 1]$ when $d \geq 5$ by a different proof.

(ii) In [40, 103], it is proved that the scaling limit (in the sense of finite-dimensional distributions) of the BTM on $\mathbb{R}$ is the FIN diffusion.

(iii) In [39], they consider a (more) general model of trapping for random walks on graphs, and give a class of possible scaling limits of the randomly trapped random walks on $\mathbb{Z}$. It turns out the class of possible limits are much richer than the FK process and the FIN diffusion.

Remark 8.1.9. (i) Isoperimetric inequalities and heat kernel estimates are very useful to obtain various properties of the random walk. For the case of supercritical percolation in a box, estimates of mixing times [49] and the Laplace transform of the range of a random walk [193] are obtained with the help of isoperimetric inequalities and heat kernel estimates.

(ii) For two-dimensional supercritical percolation [35, 71], and other models including IICs on trees and IIC on $\mathbb{Z}^d$ for d large [35], it is proved that two independent random walks started at the same point collide infinitely often $\mathbb{P}$-a.s.

8.2   Percolation Estimates

Consider the supercritical bond percolation $p > p_c(d)$ on $\mathbb{Z}^d$ for $d \geq 2$. In this section, we will give some percolation estimates that are needed later. We do not give a proof here, but mention the corresponding references.

Let $\mathcal{C}(0)$ be the open cluster containing 0 and for $x, y \in \mathcal{C}(0)$, let $d_\omega(x, y)$ be the graph distance for $\mathcal{C}(0)$ and $|x - y|$ be the Euclidean distance. The first statement of the next proposition gives a stretched-exponential decay of truncated connectivities due to [122, Theorem 8.65]. The second statement is a comparison of the graph distance and the Euclidean distance due to Antal and Pisztora [15, Corollary 1.3].

Proposition 8.2.1. *Let* $p > p_c(d)$. *Then the following hold.*

(i) *There exists* $c_1 = c_1(p)$ *such that*

$$\mathbb{P}_p(|\mathcal{C}(0)| = n) \leq \exp(-c_1 n^{(d-1)/d}) \quad \forall n \in \mathbb{N}.$$

(ii) *There exists* $c_2 = c_2(p, d) > 0$ *such that the following holds* $\mathbb{P}_p$-*almost surely,*

$$\limsup_{|y| \to \infty} \frac{d_\omega(0, y) 1_{\{0 \leftrightarrow y\}}}{|y|} \leq c_2.$$

For $\alpha > 0$, denote $\mathcal{C}_{\infty,\alpha}$ the set of sites in $\mathbb{Z}^d$ that are connected to infinity by a path whose edges satisfy $\mu_b \geq \alpha$. The following proposition is due to [59, Proposition 2.3]. Similar estimate for the size of "holes" in $\mathcal{C}_\infty$ can be found in [178, Lemma 3.1].

Proposition 8.2.2. *Assume* $p_+ = \mathbb{P}(\mu_b > 0) > p_c(d)$. *Then there exists* $c(p_+, d) > 0$ *such that if* α *satisfies*

$$\mathbb{P}(\mu_b \geq \alpha) > p_c(d) \quad and \quad \mathbb{P}(0 < \mu_b < \alpha) < c(p_+, d), \tag{8.7}$$

then $\mathcal{C}_{\infty,\alpha} \neq \emptyset$ *and* $\mathcal{C}_\infty \setminus \mathcal{C}_{\infty,\alpha}$ *has only finite components a.s.*
   Further, if $\mathcal{K}(x)$ *is the (possibly empty) component of* $\mathcal{C}_\infty \setminus \mathcal{C}_{\infty,\alpha}$ *containing* x, *then*

$$\mathbb{P}\big(x \in \mathcal{C}_\infty, \ diam \ \mathcal{K}(x) \geq n\big) \leq c_1 e^{-c_2 n}, \quad n \geq 1, \tag{8.8}$$

for some $c_1, c_2 > 0$. *Here "diam" is the diameter in Euclidean distance on* $\mathbb{Z}^d$.

8.3   Proof of Some Heat Kernel Estimates

In this section, we prove some heat kernel estimates. We regret that only Theorem 8.1.1 (ii) will be proved here.

Proof of Theorem 8.1.1 (ii). For $\kappa > 1/d$ let $\epsilon > 0$ be such that $(1 + 4d\epsilon)/d < \kappa$. Let $\mathbb{P}$ be an i.i.d. conductance law on $\{2^{-N} : N \geq 0\}^{E_d}$ such that

$$\mathbb{P}(\mu_e = 1) > p_c(d), \quad \mathbb{P}(\mu_e = 2^{-N}) = cN^{-(1+\epsilon)}, \quad \forall N \geq 1, \qquad (8.9)$$

where $c = c(\epsilon)$ is a normalizing constant. Let $\mathbf{e}_1$ denote the unit vector in the first coordinate direction. Define the scale $\ell_N = N^{(1+4d\epsilon)/d}$ and for each $x \in \mathbb{Z}^d$, let $A_N(x)$ be the event that the configuration near x, $y = x + \mathbf{e}_1$ and $z = x + 2\mathbf{e}_1$ is as follows:

(1) $\mu_{yz} = 1$ and $\mu_{xy} = 2^{-N}$, while every other bond containing y or z has $\mu_e \leq 2^{-N}$.
(2) x is connected to the boundary of the box of side length $(\log \ell_N)^2$ centered at x by bonds with conductance one.

Since bonds with $\mu_e = 1$ percolate and since $\mathbb{P}(\mu_e \leq 2^{-N}) \asymp N^{-\epsilon}$, we have

$$\mathbb{P}(A_N(x)) \geq cN^{-[1+(4d-2)\epsilon]}. \qquad (8.10)$$

Now consider a grid of vertices $\mathbb{G}_N := [-\ell_N, \ell_N]^d \cap (a_N \mathbb{Z})^d$ where $a_N := 2(\log \ell_N)^2$. Since $\{A_N(x) : x \in \mathbb{G}_N\}$ are independent, we have

$$\mathbb{P}\left(\bigcap_{x \in \mathbb{G}_N} A_N(x)^c \right) \leq \left(1 - cN^{-[1+(4d-2)\epsilon]} \right)^{|\mathbb{G}_N|}$$

$$\leq \exp\left\{ -c\left(\frac{\ell_N}{(\log \ell_N)^2} \right)^d N^{-[1+(4d-2)\epsilon]} \right\} \leq e^{-cN^\epsilon}, \qquad (8.11)$$

so using the Borel-Cantelli, $\bigcap_{x \in \mathbb{G}_N} A_N(x)^c$ occurs only for finitely many N.

By Proposition 8.2.1 (i), every connected component of side length $(\log \ell_N)^2$ in $[-\ell_N, \ell_N]^d \cap \mathbb{Z}^d$ will eventually be connected to $\mathcal{C}_\infty$ in $[-2\ell_N, 2\ell_N]^d \cap \mathbb{Z}^d$. Summarizing, there exists $N_0 = N_0(\omega)$ with $\mathbb{P}(N_0 < \infty) = 1$ such that for $N \geq N_0$, $A_N(x)$ occurs for some $x = x_N(\omega) \in [-\ell_N, \ell_N]^d \cap \mathbb{Z}^d$ that is connected to 0 by a path (say Path_N) in $[-2\ell_N, 2\ell_N]^d$, on which only the last N_0 edges (i.e. those close to the origin) may have conductance smaller than one.

Now let $N \geq N_0$ and let n be such that $2^N \leq 2n < 2^{N+1}$. Let $x_N \in [-\ell_N, \ell_N]^d \cap \mathbb{Z}^d$ be such that $A_N(x_N)$ occurs and let r_N be the length of Path_N. Let $\alpha = \alpha(\omega)$ be the minimum of μ_e for e within N_0 steps of the origin. The Markov chain moves from 0 to x_N in time r_N with probability at least $\alpha^{N_0}(2d)^{-r_N}$, and the probability of staying on the bond (y, z) for time $2n - 2r_N - 2$ is bounded independently of ω. The transitions across (x, y) cost order 2^{-N} each. Hence we have

$$P_\omega^{2n}(0, 0) \geq c\alpha^{2N_0}(2d)^{-2r_N} 2^{-2N}. \qquad (8.12)$$

By Proposition 8.2.1 (ii), we have $r_N \leq c\ell_N$ for large N. Since $n \asymp 2^N$ and $\ell_N \leq (\log n)^\kappa$, we obtain the result. $\square$

8.4 Corrector and Quenched Invariance Principle

Our goal is to prove Theorems 8.1.2 and 8.1.4 assuming the heat kernel estimates. Let us first give overview of the proof. As usual for the FCLT, we use the "corrector". Let $\varphi = \varphi_\omega : \mathbb{Z}^d \to \mathbb{R}^d$ be a harmonic map, so that $M_t = \varphi(Y_t)$ is a P_ω^0-martingale. Let I be the identity map on $\mathbb{Z}^d$. The corrector is

$$\chi(x) = (\varphi - I)(x) = \varphi(x) - x.$$

It is referred to as the "corrector" because it corrects the non-harmonicity of the position function. For simplicity, let us consider CLT (instead of FCLT) for Y. By definition, we have

$$\frac{Y_t}{t^{1/2}} = \frac{M_t}{t^{1/2}} - \frac{\chi(Y_t)}{t^{1/2}}.$$

By choosing φ suitably, the martingale CLT gives that $M_t/t^{1/2}$ converges weakly to the normal distribution. So all we need is to prove $\chi(Y_t)/t^{1/2} \to 0$. This can be done once we have (a) $P_\omega^0(|Y_t| \geq At^{1/2})$ is small and (b) $|\chi(x)|/|x| \to 0$ as $|x| \to \infty$. (a) holds by the heat kernel upper bound, so the key is to prove (b), namely sub-linearity of the corrector. Note that there maybe many global harmonic functions, so we should chose one such that (b) holds. As we will see later, we in fact prove the sub-linearity of the corrector for $\mathcal{C}_{\infty,\alpha}$.

We now discuss details. Let

$$\Omega = \begin{cases} [0, 1]^{E_d} & \text{for Case 1,} \\ [1, \infty]^{E_d} & \text{for Case 2.} \end{cases}$$

(Note that one can choose $c = 1$ in Case 1 and Case 2.) The conductances $\{\mu_e : e \in E_d\}$ are defined on $(\Omega, \mathbb{P})$ and we write $\mu_{\{x,y\}}(\omega) = \omega_{x,y}$ for the coordinate maps. Let $T_x : \Omega \to \Omega$ denote the shift by x, namely $(T_z\omega)_{xy} := \omega_{x+z,y+z}$.

The construction of the corrector is simple and robust. Let $\{Q_{x,y}(\omega) : x, y \in \mathbb{Z}^d\}$ be a sequence of non-negative random variables such that $Q_{x,y}(T_z\omega) = Q_{x+z,y+z}(\omega)$, which is stationary and ergodic. Assume that there exists $C > 0$ such that the following hold:

$$\sum_{x \in \mathbb{Z}^d} Q_{0,x}(\omega) \leq C \quad \mathbb{P}\text{-a.e. } \omega, \quad \text{and} \quad \mathbb{E}[\sum_{x \in \mathbb{Z}^d} Q_{0,x}|x|^2] < \infty. \tag{8.13}$$

The following construction, coming from periodic homogenization, is well-known for specialists (we follow Mathieu and Piatnitski [179]). We will prove it in Sect. 8.5.

Theorem 8.4.1. *There exists a function $\chi : \Omega \times \mathbb{Z}^d \to \mathbb{R}^d$ that satisfies the following (1)–(3) $\mathbb{P}$-a.e. ω.*

(1) (Cocycle property) $\chi(\omega, 0) = 0$ and, for all $x, y \in \mathbb{Z}^d$,

$$\chi(\omega, x) - \chi(\omega, y) = \chi(T_y \omega, x - y). \tag{8.14}$$

(2) (Harmonicity) $\varphi_\omega(x) := x + \chi(\omega, x)$ enjoys $\mathcal{L}_Q \varphi_\omega^j(z) = 0$ for all $1 \leq j \leq d$ and $z \in \mathbb{Z}^d$ where φ_ω^j is the j-th coordinate of φ_ω and

$$(\mathcal{L}_Q f)(x) = \sum_{y \in \mathbb{Z}^d} Q_{x,y}(\omega)(f(y) - f(x)). \tag{8.15}$$

(3) (Square integrability) There exists $C < \infty$ such that for all $x, y \in \mathbb{Z}^d$,

$$\mathbb{E}\left[|\chi(\cdot, y) - \chi(\cdot, x)|^2 \, Q_{x,y}(\cdot) \right] < C. \tag{8.16}$$

Remark 8.4.2. (i) In [61, Lemma 3.3], it is proved that for the uniform elliptic case (Case 0), i.e. $\mathbb{P}(c_1 \leq Q_{xy} \leq c_2) = 1$ for some $c_1, c_2 > 0$, the corrector is uniquely determined by the properties (1)–(3) in Theorem 8.4.1 and (3-2) in Proposition 8.4.5 below. Recently, uniqueness of the corrector is proved for the supercritical percolation cluster in [47] (see Sect. 8.10).

(ii) In [59, 179] the corrector is defined essentially on $\Omega \times B$ where $B = \{e_1, \cdots, e_d, -e_1, \cdots, -e_d\}$ is the set of unit vectors. However, one can easily extend the domain of the corrector to $\Omega \times \mathbb{Z}^d$ by using the cocycle property.

Given the construction of the corrector, we proceed as in Biskup and Prescott [59]. We first give sufficient condition for the (uniform) sub-linearity of the corrector in this general setting and then show that the assumption given for the sufficiency of the sub-linearity can be verified for Case 1 and Case 2.

We consider open clusters with respect to the percolation for $\{Q_{xy}\}_{x,y \in \mathbb{Z}^d}$. Assume that the distribution of $\{Q_{xy}\}$ is given so that there exists unique infinite open cluster $\mathbb{P}$-a.s., and denote it by $\mathcal{C}_\infty$. We also consider (random) one-parameter family of infinite sub-clusters which we denote by $\mathcal{C}_{\infty,\alpha}$, where we set $\mathcal{C}_{\infty,0} = \mathcal{C}_\infty$. We assume $\mathbb{P}(0 \in \mathcal{C}_{\infty,\alpha}) > 0$. (The concrete choice of $\mathcal{C}_{\infty,\alpha}$ will be given later.)

Let $Y = \{Y_t\}_{t \geq 0}$ be the VSRW on $\mathcal{C}_\infty$ (with base measure $\nu_x \equiv 1$ for $x \in \mathcal{C}_\infty$) that corresponds to $\mathcal{L}_Q$ in (8.15). We introduce the trace of Markov chain to $\mathcal{C}_{\infty,\alpha}$ (cf. Sect. 2.3). Define $\sigma_1, \sigma_2, \ldots$ the time intervals between successive visits of Y to $\mathcal{C}_{\infty,\alpha}$, namely, let

$$\sigma_{j+1} := \inf \{ t > 0 : Y_{\sigma_0 + \cdots + \sigma_j + t} \in \mathcal{C}_{\infty,\alpha},$$

$$Y_{\sigma_0 + \cdots + \sigma_j + s} \neq Y_{\sigma_0 + \cdots + \sigma_j} \text{ for some } s \in (0, t]. \}, \tag{8.17}$$

with $\sigma_0 = 0$. For each $x, y \in \mathcal{C}_{\infty,\alpha}$, let $\hat{Q}_{xy} := \hat{Q}_{xy}^{(\alpha)}(\omega) = P_\omega^x(Y_{\sigma_1} = y)$ and define the operator

$$\mathcal{L}_{\hat{Q}} f(x) := \sum_{y \in \mathcal{C}_{\infty,\alpha}} \hat{Q}_{xy} \left(f(y) - f(x) \right). \tag{8.18}$$

Let $\hat{Y} = \{\hat{Y}_t\}_{t \geq 0}$ be the continuous-time random walk corresponding to $\mathcal{L}_{\hat{Q}}$.

The following theorem gives sufficient condition for the sub-linearity of the corrector ψ_ω on $\mathcal{C}_{\infty,\alpha}$. We will prove it in Sect. 8.6. Let $\mathbb{P}_\alpha(\cdot) := \mathbb{P}(\cdot | 0 \in \mathcal{C}_{\infty,\alpha})$ and let $\mathbb{E}_\alpha$ be the expectation with respect to $\mathbb{P}_\alpha$.

Theorem 8.4.3. *Fix $\alpha \geq 0$ and suppose $\psi_\omega : \mathcal{C}_{\infty,\alpha} \to \mathbb{R}^d$, $\theta > 0$ and $\hat{Y}$ satisfy the following (1)–(5) for $\mathbb{P}_\alpha$-a.e. ω:*

(1) (Harmonicity) If $\varphi_\omega(x) = (\varphi_\omega^1(x), \cdots, \varphi_\omega^d(x)) := x + \psi_\omega(x)$, then $\mathcal{L}_{\hat{Q}} \varphi_\omega^j = 0$ on $\mathcal{C}_{\infty,\alpha}$ for $1 \leq j \leq d$.

(2) (Sub-linearity on average) For every $\epsilon > 0$,

$$\lim_{n \to \infty} \frac{1}{n^d} \sum_{\substack{x \in \mathcal{C}_{\infty,\alpha} \\ |x| \leq n}} 1_{\{|\psi_\omega(x)| \geq \epsilon n\}} = 0. \tag{8.19}$$

(3) (Polynomial growth) There exists $\theta > 0$ such that

$$\lim_{n \to \infty} \max_{\substack{x \in \mathcal{C}_{\infty,\alpha} \\ |x| \leq n}} \frac{|\psi_\omega(x)|}{n^\theta} = 0. \tag{8.20}$$

(4) (Diffusive upper bounds) For a deterministic sequence $b_n = o(n^2)$ and a.e. ω,

$$\sup_{n \geq 1} \max_{\substack{x \in \mathcal{C}_{\infty,\alpha} \\ |x| \leq n}} \sup_{t \geq b_n} \frac{E_\omega^x |\hat{Y}_t - x|}{\sqrt{t}} < \infty, \tag{8.21}$$

$$\sup_{n \geq 1} \max_{\substack{x \in \mathcal{C}_{\infty,\alpha} \\ |x| \leq n}} \sup_{t \geq b_n} t^{d/2} P_\omega^x(\hat{Y}_t = x) < \infty. \tag{8.22}$$

(5) (Control of big jumps) Let $\tau_n = \inf\{t \geq 0 : |\hat{Y}_t - \hat{Y}_0| \geq n\}$. There exist $c_1 = c_1(\omega) > 1$ and $N_\omega > 0$ which is finite for $\mathbb{P}_\alpha$-a.e. ω such that $|\hat{Y}_{t \wedge \tau_n} - \hat{Y}_0| \leq c_1 n$ for all $t > 0$ and $n \geq N_\omega$, P_ω^0-a.s.

Then for $\mathbb{P}_\alpha$-a.e. ω,

$$\lim_{n \to \infty} \max_{\substack{x \in \mathcal{C}_{\infty,\alpha} \\ |x| \leq n}} \frac{|\psi_\omega(x)|}{n} = 0. \tag{8.23}$$

Remark 8.4.4. While the above theorem requires various additional information to prove sub-linearity of the corrector, for two-dimensional supercritical percolation, [51, Theorem 5.1] shows sub-linearity of the corrector directly without using properties such as (8.21), (8.22). One may ask a natural question whether harmonicity is an essential ingredient or not in establishing the sub-linearity. Namely, if $g_\omega : \mathbb{Z}^d \to \mathbb{R}$ is a stationary ergodic process with cocycle property whose gradients are integrable and have expectation zero, then the question is whether the following holds or not:

$$\lim_{n\to\infty} \frac{1}{n} \max_{x\in\mathbb{Z}^d \cap [-n,n]^d} |g_\omega(x)| = 0 \quad \text{a.s.}$$

It is obviously true for $d = 1$, but there is a simple negative answer for $d \geq 2$ by Tom Liggett (see [51, Appendix B]). Let $\{g.(x)\}_{x\in\mathbb{Z}^d}$ be i.i.d. with distribution function $P(g.(x) > u) = u^{-d}$ for $u \geq 1$. Then $\{g.(x)\}_{x\in\mathbb{Z}^d}$ satisfies all the conditions above, yet $n^{-1} \max_{|x|\leq n} |f(x)|$ has a non-trivial distributional limit as $n \to \infty$.

Now we discuss how to apply the theorem for Case 1 and Case 2.

Case 1:   In this case, we define $Q_{x,y}(\omega) = \omega_{x,y}$. Then it satisfies the conditions for $\{Q_{xy}\}$ given above including (8.13). Denote by $\mathcal{L}_\omega$ the generator of VSRW (which we denote by $\{Y_t\}_{t\geq0}$), i.e.,

$$(\mathcal{L}_\omega f)(x) = \sum_{y:y\sim x} \omega_{xy}(f(y) - f(x)). \tag{8.24}$$

$\mathcal{L}_Q = \mathcal{L}_\omega$ in this case. The infinite cluster $\mathcal{C}_\infty$ is the cluster for $\{b \in E_d : \mu_b > 0\}$ (since we assumed $p_+ = \mathbb{P}(\mu_b > 0) > p_c(d)$, it is in the supercritical regime so there exists unique infinite cluster.) Fix $\alpha \geq 0$ that satisfies (8.7) and $\mathcal{C}_{\infty,\alpha}$ the cluster for $\{b \in E_d : \mu_b \geq \alpha\}$. (Again it is the supercritical percolation so there exists unique infinite cluster.) We let $\hat{Q}_{xy} := P_\omega^x(Y_{\sigma_1} = y)$. Note that although $\hat{Y}$ may jump the "holes" $\mathcal{C}_\infty \setminus \mathcal{C}_{\infty,\alpha}$, Proposition 8.2.2 shows that all jumps are finite. Let $\varphi_\omega(x) := x + \chi(\omega, x)$ where χ is the corrector on $\mathcal{C}_\infty$. Then, by the optional stopping theorem, it is easy to see that $\mathcal{L}_{\hat{Q}}\varphi_\omega = 0$ on $\mathcal{C}_{\infty,\alpha}$ (see Lemma 8.7.1 (i)).

Case 2:   First, note that if we define $Q_{x,y}(\omega)$ similarly as in Case 1, it does not necessarily satisfy (8.13). Especially when $\mathbb{E}\mu_e = \infty$, we cannot define correctors for $\{Y_t\}_{t\geq0}$ in a usual manner. The idea of [28] is to discretize $\{Y_t\}_{t\geq0}$ and construct the corrector for the discretized process $\{Y_{[t]}\}_{t\geq0}$. Let $q_t^\omega(x, y)$ be the heat kernel of $\{Y_t\}_{t\geq0}$. Note that $q_t^\omega(x, y) = P_\omega^x(Y_t = y) = q_t^\omega(y, x)$. We define

$$Q_{x,y}(\omega) = q_1^\omega(x, y), \qquad \forall x, y \in \mathbb{Z}^d.$$

Then $Q_{x,y} \le 1$ and $\sum_y Q_{x,y} = 1$. Note that in this case $\mathbb{P}(Q_{xy} > 0) > 0$ for all $x, y \in \mathbb{Z}^d$, so $\mathcal{C}_\infty = \mathbb{Z}^d$. Integrating the inequalities in Theorem 8.1.3 (b), we have $\mathbb{E}[E_\cdot^0 |Y_1|^2] = \mathbb{E}[\sum_x Q_{0x}|x|^2] \le c$, so (8.13) holds. One can easily check other conditions for $\{Q_{xy}\}$ given above. In this case, we do not need to consider $\mathcal{C}_{\infty,\alpha}$ and there is no need to take the trace process. So, $\alpha = 0$, $\hat{Q}_{x,y} = Q_{x,y}$, $\mathcal{L}_{\hat{Q}} = \mathcal{L}_Q$, and $\hat{Y}_t = Y_{[t]}$ (discrete time random walk) in this case.

The next proposition provides some additional properties of the corrector for Case 1 and Case 2. This together with Lemma 8.8.1 below verify for Case 1 and Case 2 the sufficient conditions for sub-linearity of the corrector given in Theorem 8.4.3. As mentioned above, for Case 2, we consider only $\alpha = 0$.

Proposition 8.4.5. *Let $\alpha > 0$ for Case 1 and $\alpha = 0$ for Case 2, and let χ be the corrector given in Theorem 8.4.1. Then χ satisfies (2), (3) in Theorem 8.4.1 for $\mathbb{P}_\alpha$-a.e. ω if $\mathbb{P}(0 \in \mathcal{C}_{\infty,\alpha}) > 0$. Further, it satisfies the following.*

(1) *(Polynomial growth) There exists $\theta > d$ such that the following holds $\mathbb{P}_\alpha$-a.e. ω:*

$$\lim_{n \to \infty} \max_{\substack{x \in \mathcal{C}_{\infty,\alpha} \\ |x| \le n}} \frac{|\chi(\omega, x)|}{n^\theta} = 0. \tag{8.25}$$

(2) *(Sub-linearity on average) For each $\epsilon > 0$, the following holds $\mathbb{P}_\alpha$-a.e. ω:*

$$\lim_{n \to \infty} \frac{1}{n^d} \sum_{\substack{x \in \mathcal{C}_{\infty,\alpha} \\ |x| \le n}} 1_{\{|\chi(\omega,x)| \ge \epsilon n\}} = 0.$$

(3-1) *Case 1: (Zero mean under random shifts) Let $Z: \Omega \to \mathbb{Z}^d$ be a random variable such that (a) $Z(\omega) \in \mathcal{C}_{\infty,\alpha}(\omega)$, (b) $\mathbb{P}_\alpha$ is preserved by $\omega \mapsto \tau_{Z(\omega)}(\omega)$, and (c) $\mathbb{E}_\alpha[d_\omega^{(\alpha)}(0, Z(\omega))^q] < \infty$ for some $q > 3d$. Then $\chi(\cdot, Z(\cdot)) \in L^1(\Omega, \mathcal{F}, \mathbb{P}_\alpha)$ and*

$$\mathbb{E}_\alpha[\chi(\cdot, Z(\cdot))] = 0. \tag{8.26}$$

(3-2) *Case 2: (Zero mean) $\mathbb{E}[\chi(\cdot, x)] = 0$ for all $x \in \mathbb{Z}^d$.*

In (3-1), we let $d_\omega^{(\alpha)}(x, y)$ be the graph distance between x and y on $\mathcal{C}_{\infty,\alpha}$. The proof of this proposition will be given in Sect. 8.7.

8.5   Construction of the Corrector

In this section, we prove Theorem 8.4.1. We follow the arguments in Barlow and Deuschel [28, Sect. 5] (see also [59, 61, 179]).

We define the process that gives the "environment seen from the particle" by

$$\omega_t = T_{Y_t}(\omega), \qquad \forall t \in [0, \infty), \tag{8.27}$$

where $\{Y_t\}_{t\geq 0}$ is the Markov chain corresponding to $\mathcal{L}_Q$. Note that the process $\omega_{\cdot}$ is ergodic under the time shift on Ω (see for example, [51, Sect. 3], [91, Lemma 4.9] for the proof in discrete time).

Let $\mathbb{L}^2 = L^2(\Omega, \mathbb{P})$ and for $F \in \mathbb{L}^2$, write $F_x = F \circ T_x$. Then the generator of $\omega_{\cdot}$ is

$$\hat{L}F(\omega) = \sum_{x \in \mathbb{Z}^d} Q_{0,x}(\omega)(F_x(\omega) - F(\omega)). \tag{8.28}$$

Define

$$\hat{\mathcal{E}}(F, G) = \mathbb{E}\left[\sum_{x \in \mathbb{Z}^d} Q_{0,x}(F - F_x)(G - G_x) \right] \qquad \forall F, G \in \mathbb{L}^2.$$

The following lemma shows that $\hat{\mathcal{E}}$ is the quadratic form on $\mathbb{L}^2$ corresponding to $\hat{L}$.

Lemma 8.5.1. *(i) For all $F \in \mathbb{L}^2$ and $x \in \mathbb{Z}^d$, it holds that $\mathbb{E}F = \mathbb{E}F_x$ and $\mathbb{E}[Q_{0,x}F_x] = \mathbb{E}[Q_{0,-x}F]$.*
(ii) For $F \in \mathbb{L}^2$, it holds that $\hat{\mathcal{E}}(F, F) < \infty$ and $\hat{L}F \in \mathbb{L}^2$.
(iii) For $F, G \in \mathbb{L}^2$, it holds that $\hat{\mathcal{E}}(F, G) = -\mathbb{E}[G\hat{L}F]$.

Proof. (i) The first equality is because $\mathbb{P} = \mathbb{P} \circ T_x$. Since $Q_{0,x} \circ T_{-x} = Q_{-x,0} = Q_{0,-x}$, $\mathbb{E}[Q_{0,x}F_x] = \mathbb{E}[(Q_{0,x} \circ T_{-x})F] = \mathbb{E}[Q_{0,-x}F]$ so the second equality holds.

(ii) For $F \in \mathbb{L}^2$, we have

$$\hat{\mathcal{E}}(F, F) = \mathbb{E}[\sum_x Q_{0,x}(F - F_x)^2] \leq 2\mathbb{E}[\sum_x Q_{0,x}(F^2 + F_x^2)]$$

$$= 2C\mathbb{E}F^2 + 2\mathbb{E}[\sum_x Q_{0,-x}F^2] = 4C\|F\|_2^2, \tag{8.29}$$

where we used (i) in the second equality, and C is the constant in (8.13). Also,

$$\mathbb{E}|\hat{L}F|^2 = \mathbb{E}[\sum_{x,y} Q_{0,x}Q_{0,y}(F_x - F)(F_y - F)]$$

$$\leq \mathbb{E}\left[(\sum_{x,y} Q_{0,x}Q_{0,y}(F_x - F)^2)^{1/2}(\sum_{x,y} Q_{0,x}Q_{0,y}(F_y - F)^2)^{1/2} \right]$$

$$\leq C\hat{\mathcal{E}}(F, F) \leq 4C^2\|F\|_2^2$$

where (8.29) is used in the last inequality. We thus obtain (ii).

(iii) Using (i), we have

$$\mathbb{E}[Q_{0,-x}G(F - F_{-x})] = \mathbb{E}[Q_{0,x}G_x(F_x - F)]. \qquad (8.30)$$

So

$$-\mathbb{E}[G\hat{L}F] = \sum_x \mathbb{E}[GQ_{0,x}(F - F_x)]$$

$$= \frac{1}{2}\sum_x \mathbb{E}[GQ_{0,x}(F - F_x)] + \frac{1}{2}\sum_x \mathbb{E}[GQ_{0,-x}(F - F_{-x})]$$

$$= \frac{1}{2}\sum_x \mathbb{E}[Q_{0,x}(GF - GF_x + G_xF_x - G_xF)] = \hat{\mathcal{E}}(F, G),$$

where (8.30) is used in the third equation, and (iii) is proved.   □

Next we look at "vector fields". Let M be the measure on $\Omega \times \mathbb{Z}^d$ defined as

$$\int_{\Omega \times \mathbb{Z}^d} G dM := \mathbb{E}\Big[\sum_{x \in \mathbb{Z}^d} Q_{0,x}G(\cdot, x)\Big], \quad \text{for } G : \Omega \times \mathbb{Z}^d \to \mathbb{R}.$$

We say $G : \Omega \times \mathbb{Z}^d \to \mathbb{R}$ has the *cocycle property* (or *shift-covariance property*) if the following holds,

$$G(T_x\omega, y - x) = G(\omega, y) - G(\omega, x), \quad \mathbb{P}\text{-a.s.}$$

Let $\overline{L}^2 = \{G \in L^2(\Omega \times \mathbb{Z}^d, M) : G$ has the cocycle property.$\}$ and for $G \in L^2(\Omega \times \mathbb{Z}^d, M)$, write $\|G\|_{\overline{L}^2}^2 := \mathbb{E}[\sum_{x \in \mathbb{Z}^d} Q_{0,x}G(\cdot, x)^2]$. It is easy to see that $\overline{L}^2$ is a Hilbert space. Also, for $G \in \overline{L}^2$, it holds that $G(\omega, 0) = 0$ and $G(T_x\omega, -x) = -G(\omega, x)$.

Define $\nabla : L^2 \to \overline{L}^2$ by

$$\nabla F(\omega, x) := F(T_x\omega) - F(\omega), \quad \text{for } F \in L^2.$$

Since $\|\nabla F\|_{\overline{L}^2}^2 = \hat{\mathcal{E}}(F, F) < \infty$ (due to Lemma 8.5.1 (ii)) and $\nabla F(T_x\omega, y - x) = F(T_y\omega) - F(T_x\omega) = \nabla F(\omega, y) - \nabla F(\omega, x)$, we can see that ∇F is indeed in $\overline{L}^2$.

Now we introduce an orthogonal decomposition of $\overline{L}^2$ as follows

$$\overline{L}^2 = L_{\text{pot}}^2 \oplus L_{\text{sol}}^2.$$

Here $L_{\text{pot}}^2 = Cl\{\nabla F : F \in L^2\}$ where the closure is taken in $\overline{L}^2$, and L_{sol}^2 is the orthogonal complement of L_{pot}^2. Note that "pot" stands for "potential", and "sol" stands for "solenoidal".

Before giving some properties of L_{pot}^2 and L_{sol}^2, we give definition of the corrector. Let $\Pi : \mathbb{R}^d \to \mathbb{R}^d$ be the identity and denote by Π_j the j-th coordinate of Π. Then, $\Pi_j \in \overline{L}^2$. Indeed, $\Pi_j(y-x) = \Pi_j(y) - \Pi_j(x)$ so it has the cocycle property, and by (8.13), $\|\Pi_j\|_{\overline{L}^2}^2 < \infty$. Now define $\chi_j \in L_{\text{pot}}^2$ and $\varphi_j \in L_{\text{sol}}^2$ by

$$\Pi_j = (-\chi_j) \oplus \varphi_j \in L_{\text{pot}}^2 \oplus L_{\text{sol}}^2.$$

This gives the definition of the corrector $\chi = (\chi_1, \cdots, \chi_d) : \Omega \times \mathbb{Z}^d \to \mathbb{R}^d$.

Remark 8.5.2. The corrector can be defined by using spectral theory as in Kipnis and Varadhan [154]. This "projection" definition is given in Mathieu and Piatnitski [179] (also in [28, 59, 61, 111] etc.). There is yet another definition of the corrector as a limit point of a "harness process" (see [100]).

Lemma 8.5.3. *For $G \in L_{\text{sol}}^2$, it holds that*

$$\sum_{x \in \mathbb{Z}^d} Q_{0,x} G(\omega, x) = 0, \quad \mathbb{P}\text{-}a.s. \tag{8.31}$$

Hence $M_n := G(\omega, Y_n)$ is a P_ω^0-martingale for $\mathbb{P}$-a.e. ω.

Proof. Recall that $G(T_x\omega, -x) = -G(\omega, x)$ for $G \in \overline{L}^2$. Using this, we have for each $F \in \mathbb{L}^2$

$$\sum_x \mathbb{E}[Q_{0,x} G(\cdot, x) F_x(\cdot)] = \sum_x \mathbb{E}[T_{-x} Q_{0,x} G(T_{-x}\cdot, x) F_x(T_{-x}\cdot)]$$

$$= \sum_x \mathbb{E}[Q_{0,-x}(-G(\cdot, -x)) F(\cdot)]$$

$$= -\sum_x \mathbb{E}[Q_{0,x} G(\cdot, x) F(\cdot)].$$

So $\sum_x \mathbb{E}[Q_{0,x} G(\cdot, x)(F + F_x)(\cdot)] = 0$. If $G \in L_{\text{sol}}^2$, then since $\nabla F \in L_{\text{pot}}^2$, we have

$$0 = \int_{\Omega \times \mathbb{Z}^d} G \nabla F dM = \sum_x \mathbb{E}[Q_{0,x} G(\cdot, x)(F_x - F)].$$

So we have $\mathbb{E}[\sum_x Q_{0,x} GF] = 0$. Since this holds for all $F \in \mathbb{L}^2$, we obtain (8.31). Since

$$E_\omega^0[G(\omega, Y_{n+1}) - G(\omega, Y_n)|Y_n = x] = \sum_y Q_{xy}(\omega)(G(\omega, y) - G(\omega, x))$$

$$= \sum_y Q_{0,y-x}(T_x\omega) G(T_x\omega, y-x) = 0,$$

where the cocycle property is used in the second equality and (8.31) is used in the last inequality, we see that $M_n := G(\omega, Y_n)$ is a P_ω^0-martingale. □

Verification of (1)–(3) in Theorem 8.4.1: (1) and (3) in Theorem 8.4.1 is clear by definition of χ and the definition of the inner product on $\mathbb{L}^2(\Omega \times \mathbb{Z}^d, M)$. By Lemma 8.5.3, we see that (2) in Theorem 8.4.1 hold. Here, note that by the cocycle property of φ,

$$(\mathcal{L}_Q \varphi_\omega^j)(x) = \sum_{y \in \mathbb{Z}^d} Q_{x,y}(\omega)(\varphi_\omega^j(y) - \varphi_\omega^j(x)) = \sum_{y \in \mathbb{Z}^d} Q_{0,y-x}(T_x\omega)\varphi_{T_x\omega}^j(y-x) = 0,$$

for $1 \leq j \leq d$ where the last equality is due to (8.31). □

In [61, Sect. 5], there is a nice survey of the potential theory behind the notion of the corrector. There it is shown that $\Pi - \chi$ (here Π is the identity map) spans L_{sol}^2 (see [61, Corollary 5.6]).

8.6 Proof of Theorem 8.4.3

The basic idea of the proof of Theorem 8.4.3 is that sub-linearity on average plus heat kernel upper bounds imply pointwise sub-linearity. In Theorem 8.4.3, the inputs that should come from the heat kernel upper bounds are reduced to the assumptions (8.21), (8.22). In order to prove Theorem 8.4.3, the next lemma plays a key role.

Lemma 8.6.1. *Let*

$$\bar{R}_n := \max_{\substack{x \in \mathcal{C}_{\infty,\alpha} \\ |x| \leq n}} |\psi_\omega(x)|.$$

Under the conditions (1,2,4,5) of Theorem 8.4.3, for each $\epsilon > 0$ and $\delta > 0$, there exists $c_1 = c_1(\omega) > 1$ and a random variable $n_0 = n_0(\omega, \epsilon, \delta)$ which is a.s. finite such that

$$\bar{R}_n \leq \epsilon n + \delta \bar{R}_{2c_1 n}. \qquad n \geq n_0. \tag{8.32}$$

Before proving this lemma, let us prove Theorem 8.4.3 by showing how this lemma and (8.20) imply (8.23).

Proof of Theorem 8.4.3. Suppose that $\bar{R}_n/n \not\to 0$ and choose c with $0 < c < \limsup_{n\to\infty} \bar{R}_n/n$. Let θ be is as in (8.20) and choose $\epsilon := c/2$ and $\delta := (2c_1)^{-\theta-1}$. Note that then $c - \epsilon \geq (2c_1)^\theta \delta c$. If $\bar{R}_n \geq cn$ (this happens for infinitely many n's) and $n \geq n_0$, then (8.32) implies

$$\bar{R}_{2c_1 n} \geq \frac{c-\epsilon}{\delta}n \geq (2c_1)^\theta cn$$

and, inductively, $\bar{R}_{(2c_1)^k n} \geq (2c_1)^{k\theta} cn$. However, (8.20) says $\bar{R}_{(2c_1)^k n}/(2c_1)^{k\theta} \to 0$ as $k \to \infty$ for each fixed n, which is a contradiction. $\square$

We now prove Lemma 8.6.1 following [51]. The idea behind is the following. Let $\{\hat{Y}_t\}$ be the process corresponding to $\mathcal{L}_{\hat{Q}}$ started at the maximizer of $\bar{R}_n$, say z^*. Using the martingale property, we will estimate $E_\omega^{z^*}[\psi_\omega(\hat{Y}_t)]$ for time $t = o(n^2)$. The right-hand side of (8.32) expresses two situations that may occur at time t: (i) $|\psi_\omega(\hat{Y}_t)| \leq \epsilon n$ (by "sub-linearity on average", this happens with high probability), (ii) $\hat{Y}$ has not yet left the box $[-2c_1 n, 2c_1 n]^d$ and so $\psi_\omega(\hat{Y}_t) \leq \bar{R}_{2c_1 n}$. It turns out that these two are the dominating strategies.

Proof of Lemma 8.6.1. Fix $\epsilon, \delta > 0$ and let z_* be the site where the maximum $\bar{R}_n$ is achieved. Denote

$$\mathcal{O}_n := \left\{ x \in \mathcal{C}_{\infty,\alpha} : |x - z_*| \leq n, \ |\psi_\omega(x)| \geq \tfrac{1}{2}\epsilon n \right\}.$$

Recall that $\{\hat{Y}_t\}$ is the continuous-time random walk on $\mathcal{C}_{\infty,\alpha}$ corresponding to $\mathcal{L}_{\hat{Q}}$ in (8.18). We denote its expectation for the walk started at z_* by $E_\omega^{z_*}$. Define

$$S_n := \inf\{t \geq 0 : |\hat{Y}_t - z_*| \geq 2n\}.$$

Note that, by Theorem 8.4.3 (5), there exists $n_1(\omega)$ which is a.s. finite such that we have $|\hat{Y}_{t \wedge S_n} - z_*| \leq 2c_1 n$ for all $t > 0$ and $n \geq n_1(\omega)$. Using the harmonicity of $x \mapsto x + \psi_\omega(x)$ and the optional stopping theorem, we have

$$\bar{R}_n = |\psi_\omega(z_*)| \leq E_\omega^{z_*}\left[\left| \psi_\omega(\hat{Y}_{t \wedge S_n}) + \hat{Y}_{t \wedge S_n} - z_* \right| \right]. \tag{8.33}$$

We will consider t that satisfies

$$t \geq b_{3c_1 n}, \tag{8.34}$$

where $b_n = o(n^2)$ is the sequence in Theorem 8.4.3 (4), and we will estimate the expectation separately on $\{S_n < t\}$ and $\{S_n \geq t\}$.

Case i $\{S_n < t\}$: On this event, the integrand of (8.33)$\leq \bar{R}_{2c_1 n} + 2c_1 n$. To estimate $P_{\omega, z_*}(S_n < t)$, we consider the subcases $|\hat{Y}_{2t} - z_*| \geq n$ and $\leq n$. For the former case, (8.34) and (8.21) imply

$$P_{\omega, z_*}\left(|\hat{Y}_{2t} - z_*| \geq n \right) \leq \frac{E_\omega^{z_*}|\hat{Y}_{2t} - z_*|}{n} \leq \frac{c_2 \sqrt{t}}{n},$$

for some $c_2 = c_2(\omega)$. For the latter case, we have

$$\{|\hat{Y}_{2t} - z_*| \leq n\} \cap \{S_n < t\} \subset \{|\hat{Y}_{2t} - \hat{Y}_{S_n}| \geq n\} \cap \{S_n < t\}.$$

Noting that $2t - S_n \in [t, 2t]$ and using (8.34), (8.21) again, we have

$$P_{\omega,x}\left(|\hat{Y}_s - x| \geq n\right) \leq \frac{c_3\sqrt{t}}{n} \qquad \text{when } x := \hat{Y}_{S_n} \text{ and } s := 2t - S_n,$$

for some $c_3 = c_3(\omega)$. From the strong Markov property we have the same bound for $P_{\omega,z_*}(S_n < t, |\hat{Y}_{2t} - z_*| \geq n)$. Combining both cases, we have $P_{\omega,z_*}(S_n < t) \leq (c_2 + c_3)\sqrt{t}/n$. So

$$E_{\omega}^{z_*}\left[|\psi_{\omega}(\hat{Y}_{t \wedge S_n}) + \hat{Y}_{t \wedge S_n} - z_*|1_{\{S_n < t\}}\right] \leq (\bar{R}_{2c_1n} + 2c_1n)\frac{(c_2 + c_3)\sqrt{t}}{n}. \qquad (8.35)$$

Case ii $\{S_n \geq t\}$: On this event, the expectation in (8.33) is bounded by

$$E_{\omega}^{z_*}\left[|\psi_{\omega}(\hat{Y}_t)|1_{\{S_n \geq t\}}\right] + E_{\omega}^{z_*}|\hat{Y}_t - z_*|.$$

By (8.21), the second term is less than $c_4\sqrt{t}$ for some $c_4 = c_4(\omega)$ provided $t \geq b_n$. The first term can be estimated depending on whether $\hat{Y}_t \notin \mathcal{O}_{2n}$ or not:

$$E_{\omega}^{z_*}\left[|\psi_{\omega}(\hat{Y}_t)|1_{\{S_n \geq t\}}\right] \leq \epsilon n + \bar{R}_{2c_1n}P_{\omega,z_*}(\hat{Y}_t \in \mathcal{O}_{2n}).$$

For the probability of $\hat{Y}_t \in \mathcal{O}_{2n}$ we get

$$P_{\omega,z_*}(\hat{Y}_t \in \mathcal{O}_{2n}) = \sum_{x \in \mathcal{O}_{2n}} P_{\omega,z_*}(\hat{Y}_t = x)$$

$$\leq \sum_{x \in \mathcal{O}_{2n}} P_{\omega,z_*}(\hat{Y}_t = z_*)^{1/2} P_{\omega,x}(\hat{Y}_t = x)^{1/2} \leq c_5 \frac{|\mathcal{O}_{2n}|}{t^{d/2}},$$

where $c_5 = c_5(\omega)$ denote the suprema in (8.22). Here is the first inequality we used the Schwarz inequality and the second inequality is due to (8.22).

$$E_{\omega}^{z_*}\left[|\psi_{\omega}(\hat{Y}_{t \wedge S_n}) + \hat{Y}_{t \wedge S_n} - z_*|1_{\{S_n \geq t\}}\right] \leq c_4\sqrt{t} + \epsilon n + \bar{R}_{2c_1n}c_5\frac{|\mathcal{O}_{2n}|}{t^{d/2}}. \qquad (8.36)$$

By (8.35) and (8.36), we conclude that

$$\bar{R}_n \leq (\bar{R}_{2c_1n} + 2c_1n)\frac{(c_2 + c_3)\sqrt{t}}{n} + c_4\sqrt{t} + \epsilon n + \bar{R}_{2c_1n}c_5\frac{|\mathcal{O}_{2n}|}{t^{d/2}}. \qquad (8.37)$$

Since $|\mathcal{O}_{2n}| = o(n^d)$ as $n \to \infty$ due to (8.19), we can choose $t := \xi n^2$ with $\xi > 0$ so small that (8.34) applies and (8.32) holds for the given ϵ and δ once n is sufficiently large. $\qquad \square$

8.7   Proof of Proposition 8.4.5

In this section, we prove Proposition 8.4.5 for Case 1 and Case 2 separately.

8.7.1   Case 1

For Case 1, we need to work on $C_{\infty,\alpha}$, but we do not need to use Proposition 8.2.2 seriously yet (although we use it lightly in (8.42)). By definition, it is clear that χ satisfies Theorem 8.4.1 (3) for $\mathbb{P}_\alpha$ for all $\alpha \geq 0$ with $\mathbb{P}(0 \in C_{\infty,\alpha}) > 0$. The next lemma proves Theorem 8.4.1 (2) for $\mathbb{P}_\alpha$.

Lemma 8.7.1. *Let χ be the corrector on C_∞ and define $\varphi_\omega(x) = (\varphi_\omega^1(x), \cdots, \varphi_\omega^d(x)) := x + \chi(\omega, x)$. Then $\mathcal{L}_{\hat{Q}} \varphi_\omega^j(x) = 0$ for all $x \in C_{\infty,\alpha}$ and $1 \leq j \leq d$.*

Proof. Note that $(\mathcal{L}_{\hat{Q}}\varphi_\omega)(x) = E_\omega^x[\varphi_\omega(Y_{\sigma_1})] - \varphi_\omega(x)$. Also, since $\{Y_t\}_{t \geq 0}$ moves in a finite component of $C_\infty \setminus C_{\infty,\alpha}$ for $t \in [0, \sigma_1]$, $\varphi_\omega(Y_t)$ is bounded. Since $\{\varphi_\omega(Y_t)\}_{t \geq 0}$ is a martingale and $\sigma_1 < \infty$ a.s., the optional stopping theorem gives $E_\omega^x \varphi_\omega(Y_{\sigma_1}) = \varphi_\omega(x)$. $\square$

Proof of Proposition 8.4.5 (1). In Case 1, the result holds for all $\theta > d$ as we shall prove. Let $\theta > d$ and write

$$R_n := \max_{\substack{x \in C_{\infty,\alpha} \\ |x| \leq n}} |\chi(\omega, x)|.$$

By Proposition 8.2.1 (ii),

$$\lambda(\omega) := \sup_{x \in C_{\infty,\alpha}} \frac{d_\omega^{(\alpha)}(0, x)}{|x|} < \infty, \qquad \mathbb{P}_\alpha\text{-a.s.} \tag{8.38}$$

Recall that $d_\omega^{(\alpha)}(x, y)$ is the graph distance between x and y on $C_{\infty,\alpha}$. So it is enough to prove $R_n/n^\theta \to 0$ on $\{\lambda(\omega) \leq \lambda\}$ for all $\lambda < \infty$. Note that on $\{\lambda(\omega) \leq \lambda\}$ every $x \in C_{\infty,\alpha} \cap [-n,n]^d$ can be connected to the origin by a path that is inside $C_{\infty,\alpha} \cap [-\lambda n, \lambda n]^d$. Using this fact and $\chi(\omega, 0) = 0$, we have, on $\{\lambda(\omega) \leq \lambda\}$,

$$R_n \leq \sum_{\substack{x \in C_{\infty,\alpha} \\ |x| \leq \lambda n}} \sum_{b \in B} 1_{\{\omega_{x,x+b} \in C_{\infty,\alpha}\}} |\chi(\omega, x + b) - \chi(\omega, x)|$$

$$\leq \sum_{\substack{x \in C_{\infty,\alpha} \\ |x| \leq \lambda n}} \sum_{b \in B} \sqrt{\frac{\omega_{x,x+b}}{\alpha}} |\chi(\omega, x + b) - \chi(\omega, x)|, \tag{8.39}$$

where $B = \{\mathbf{e}_1, \cdots, \mathbf{e}_d, -\mathbf{e}_1, \cdots, -\mathbf{e}_d\}$ is the set of unit vectors. Using the Schwarz and (8.16), we get

$$\mathbb{E}_\alpha[(R_n 1_{\{\lambda(\omega) \leq \lambda\}})^2] \leq \frac{2d(\lambda n)^d}{\alpha} \mathbb{E}_\alpha\Big[\sum_{\substack{x \in \mathcal{C}_{\infty,\alpha} \\ |x| \leq \lambda n}} \sum_{b \in B} \omega_{x,x+b} |\chi(\omega, x+b) - \chi(\omega, x)|^2 \Big]$$

$$\leq C n^{2d} \tag{8.40}$$

for some constant $C = C(\alpha, \lambda, d) < \infty$. Applying the Chebyshev inequality,

$$\mathbb{P}_\alpha\big(R_n 1_{\{\lambda(\omega) \leq \lambda\}} \geq n^\theta\big) \leq \frac{C}{n^{2(\theta-d)}}.$$

Taking $n = 2^k$ and using the Borel-Cantelli, we have $R_{2^k}/2^{k\theta} \leq C$ on $\{\lambda(\omega) \leq \lambda\}$ for large k, so $R_{2^k}/2^{k\theta} \to 0$ a.e. Since $R_{2^k}/2^{k+1} \leq R_n/n \leq 2R_{2^{k+1}}/2^{k+1}$ for $2^k \leq n \leq 2^{K+1}$, we have $R_n/n^\theta \to 0$ a.s. on $\{\lambda(\omega) \leq \lambda\}$. $\qquad\square$

We will need the following lemma which is easy to prove using the Hölder inequality (see [51, Lemma 4.5] for the proof).

Lemma 8.7.2. *Let $p > 1$ and $r \in [1, p)$. Let $\{X_n\}_{n \in \mathbb{N}}$ be random variables such that $\sup_{j \geq 1} \|X_j\|_p < \infty$ and let N be a random variable taking values in positive integers such that $N \in L^s$ for some $s > r(1 + \frac{1}{p})/(1 - \frac{r}{p})$. Then $\sum_{j=1}^N X_j \in L^r$. Explicitly, there exists $C = C(p, r, s) > 0$ such that*

$$\Big\| \sum_{j=1}^N X_j \Big\|_r \leq C \big(\sup_{j \geq 1} \|X_j\|_p \big) \big(\|N\|_s \big)^{s[\frac{1}{r} - \frac{1}{p}]},$$

where C is a finite constant depending only on p, r and s.

Proof of Proposition 8.4.5 (3-1). Let Z be a random variable satisfying the properties (a)–(c). Since $\chi \in L_{\text{pot}}^2$, there exists a sequence $\psi_n \in \mathbb{L}^2$ such that

$$\chi_n(\cdot, x) := \psi_n \circ T_x - \psi_n \underset{n \to \infty}{\longrightarrow} \chi(\cdot, x) \qquad \text{in } \mathbb{L}^2(\Omega \times \mathbb{Z}^d).$$

For notational simplicity assume that $\chi_n(\cdot, x) \to \chi(\cdot, x)$ almost surely. Since Z is $\mathbb{P}_\alpha$-preserving, we have $\mathbb{E}_\alpha[\chi_n(\cdot, Z(\cdot))] = 0$ once we can show that $\chi_n(\cdot, Z(\cdot)) \in \mathbb{L}^1$. It thus suffices to prove that

$$\chi_n(\cdot, Z(\cdot)) \underset{n \to \infty}{\longrightarrow} \chi(\cdot, Z(\cdot)) \qquad \text{in } \mathbb{L}^1. \tag{8.41}$$

Abbreviate $K(\omega) := d_\omega^{(\alpha)}(0, Z(\omega))$ and note that, as in (8.39),

$$\left|\chi_n(\omega, Z(\omega))\right| \leq \sum_{\substack{x \in \mathcal{C}_{\infty,\alpha} \\ |x| \leq K(\omega)}} \sum_{b \in B} \sqrt{\frac{\omega_{x,x+b}}{\alpha}} \left|\chi_n(\omega, x+b) - \chi_n(\omega, x)\right|.$$

Note that $\sqrt{\omega_{x,x+b}} \left|\chi_n(\omega, x+b) - \chi_n(\omega, x)\right| 1_{\{x \in \mathcal{C}_{\infty,\alpha}\}}$ is bounded in $\mathbb{L}^2$, uniformly in x, b and n, and assumption (c) shows that $K \in \mathbb{L}^q$ for some $q > 3d$. Thus, by choosing $p = 2$, $s = q/d$ and $N = 2d(2K+1)^d$ in Lemma 8.7.2 (note that $N \in \mathbb{L}^s$), we obtain,

$$\sup_{n \geq 1} \|\chi_n(\cdot, Z(\cdot))\|_r < \infty,$$

for some $r > 1$. Hence, the family $\{\chi_n(\cdot, Z(\cdot))\}$ is uniformly integrable, so (8.41) follows by the fact that $\chi_n(\cdot, Z(\cdot))$ converge almost surely. $\square$

The proof of Proposition 8.4.5 (2) is quite involved since there are random holes in $\mathcal{C}_{\infty,\alpha}$. We follow the approach in Berger and Biskup [51].

Lemma 8.7.3. *For* $\omega \in \{0 \in \mathcal{C}_{\infty,\alpha}\}$, *let* $\{x_n(\omega)\}_{n \in \mathbb{Z}}$ *be the intersections of* $\mathcal{C}_{\infty,\alpha}$ *and one of the coordinate axis so that* $x_0(\omega) = 0$. *Then*

$$\lim_{n \to \infty} \frac{\chi(\omega, x_n(\omega))}{n} = 0, \qquad \mathbb{P}_\alpha\text{-}a.s.$$

Proof. Similarly to (8.27), let $\sigma(\omega) := T_{x_1(\omega)}(\omega)$ denote the "induced" shift. As before, it is standard to prove that σ is $\mathbb{P}_\alpha$-preserving and ergodic (cf. [51, Theorem 3.2]). Further, using Proposition 8.2.2,

$$\mathbb{E}_\alpha\left(d_\omega^{(\alpha)}(0, x_1(\omega))^p\right) < \infty, \qquad \forall p < \infty. \tag{8.42}$$

Now define $\Psi(\omega) := \chi(\omega, x_1(\omega))$. Using (8.42), we can use Proposition 8.4.5 (3-1) with $Z(\omega) = x_1(\omega)$ and obtain

$$\Psi \in \mathbb{L}^1(\Omega, \mathbb{P}_\alpha) \quad \text{and} \quad \mathbb{E}_\alpha \Psi(\omega) = 0.$$

Using the cocycle property of χ, we can write

$$\frac{\chi(\omega, x_n(\omega))}{n} = \frac{1}{n} \sum_{k=0}^{n-1} \Psi \circ \sigma^k(\omega)$$

and so the left-hand side tends to zero $\mathbb{P}_\alpha$-a.s. by Birkhoff's ergodic theorem. $\square$

Now we should "build up" averages over higher dimensional boxes inductively, which is the next lemma. We first give some definition and notation.

Given $K > 0$ and $\epsilon > 0$, we say that a site $x \in \mathbb{Z}^d$ is (K, ϵ)-*good* if $x \in C_\infty(\omega)$ and

$$\left| \chi(y, \omega) - \chi(x, \omega) \right| < K + \epsilon |x - y|$$

holds for every $y \in C_{\infty,\alpha}(\omega)$ of the form $y = \ell \mathbf{e}$, where $\ell \in \mathbb{Z}$ and $\mathbf{e}$ is a unit vector. We denote by $\mathcal{G}_{K,\epsilon} = \mathcal{G}_{K,\epsilon}(\omega)$ the set of (K, ϵ)-good sites.

For each $v = 1, \ldots, d$, let Λ_n^v be the v-dimensional box

$$\Lambda_n^v = \left\{ k_1 e_1 + \cdots + k_v e_v : k_i \in \mathbb{Z}, \; |k_i| \leq n, \; \forall i = 1, \ldots, v \right\}.$$

For each $\omega \in \Omega$, define

$$\varrho_v(\omega) = \lim_{\epsilon \downarrow 0} \lim_{n \to \infty} \sup \inf_{y \in C_{\infty,\alpha}(\omega) \cap \Lambda_n^1} \frac{1}{|\Lambda_n^v|} \sum_{x \in C_{\infty,\alpha}(\omega) \cap \Lambda_n^v} 1_{\{|\chi(\omega,x) - \chi(\omega,y)| \geq \epsilon n\}}. \quad (8.43)$$

Note that the infimum is taken only over sites in one-dimensional box Λ_n^1. The idea is to prove $\varrho_v = 0$ $\mathbb{P}_\alpha$-a.s. inductively for all $1 \leq v \leq d$.

Our goal is to show by induction that $\varrho_v = 0$ almost surely for all $v = 1, \ldots, d$. The induction step is given the following lemma which is due to [51, Lemma 5.5]. (Note that a slightly more explicit version of this argument is given in [55, Sect. 4.3].)

Lemma 8.7.4. *Let* $1 \leq v < d$. *If* $\varrho_v = 0$, $\mathbb{P}_\alpha$-*almost surely, then also* $\varrho_{v+1} = 0$, $\mathbb{P}_\alpha$-*almost surely.*

Proof. Let $p_\infty = \mathbb{P}_\alpha(0 \in C_{\infty,\alpha})$. Let $v < d$ and suppose that $\varrho_v = 0$, $\mathbb{P}_\alpha$-a.s. Fix δ with $0 < \delta < \frac{1}{2} p_\infty^2$. Consider the collection of v-boxes

$$\Lambda_{n,j}^v = \tau_{j e_{v+1}}(\Lambda_n^v), \qquad j = 1, \ldots, L.$$

Here L is a deterministic number chosen so that

$$\Delta_0 = \left\{ x \in \Lambda_n^v : \exists j \in \{0, \ldots, L-1\}, \; x + j e_{v+1} \in \Lambda_{n,j}^v \cap C_{\infty,\alpha} \right\}$$

is so large that $|\Delta_0| \geq (1 - \delta)|\Lambda_n^v|$ once n is sufficiently large.

Choose $\epsilon > 0$ so that

$$L\epsilon + \delta < \frac{1}{2} p_\infty^2. \quad (8.44)$$

Pick $\epsilon > 0$ and a large value $K > 0$. Then for $\mathbb{P}_\alpha$-a.e. ω, all sufficiently large n, and for each $j = 1, \ldots, L$, there exists a set of sites $\Delta_j \subset \Lambda_{n,j}^v \cap C_{\infty,\alpha}$ such that

$$\left| (\Lambda_{n,j}^v \cap C_{\infty,\alpha}) \setminus \Delta_j \right| \leq \epsilon |\Lambda_{n,j}^v|$$

and

$$\left|\chi(x,\omega) - \chi(y,\omega)\right| \le \epsilon n, \qquad x,y \in \Delta_j. \tag{8.45}$$

Moreover, for n sufficiently large, Δ_j could be picked so that $\Delta_j \cap \Lambda_n^1 \ne \emptyset$ and, assuming K large, the non-(K,ϵ)-good sites could be pitched out with little loss of density to achieve even $\Delta_j \subset \mathcal{G}_{K,\epsilon}$. (These claims can be proved directly by the ergodic theorem and the fact that $\mathbb{P}_\alpha(0 \in \mathcal{G}_{K,\epsilon})$ converges to the density of $\mathcal{C}_{\infty,\alpha}$ as $K \to \infty$.) Given $\Delta_1, \ldots, \Delta_L$, let Λ be the set of sites in $\Lambda_n^{\nu+1} \cap \mathcal{C}_{\infty,\alpha}$ whose projection onto the linear subspace $\mathbb{H} = \{k_1 \mathbf{e}_1 + \cdots + k_\nu \mathbf{e}_\nu \colon k_i \in \mathbb{Z}\}$ belongs to the corresponding projection of $\Delta_1 \cup \cdots \cup \Delta_L$. Note that the Δ_j could be chosen so that $\Lambda \cap \Lambda_n^1 \ne \emptyset$.

By the construction, the projections of the Δ_j's, $j = 1, \ldots, L$, onto $\mathbb{H}$ "fail to cover" at most $L\epsilon|\Lambda_n^\nu|$ sites in Δ_0, and so at most $(\delta + L\epsilon)|\Lambda_n^\nu|$ sites in Λ_n^ν are not of the form $x + i\mathbf{e}_{\nu+1}$ for some $x \in \bigcup_j \Delta_j$. It follows that

$$\left|(\Lambda_n^{\nu+1} \cap \mathcal{C}_{\infty,\alpha}) \setminus \Lambda\right| \le (\delta + L\epsilon)|\Lambda_n^{\nu+1}|, \tag{8.46}$$

i.e., Λ contains all except at most $(L\epsilon + \delta)$-fraction of all sites in $\Lambda_n^{\nu+1}$. Next note that if K is sufficiently large, then for all $1 \le i < j \le L$, $\mathbb{H}$ contains at least $\frac{1}{2}p_\infty^2$-fraction of sites x such that

$$z_i := x + i\mathbf{e}_\nu \in \mathcal{G}_{K,\epsilon} \quad \text{and} \quad z_j := x + j\mathbf{e}_\nu \in \mathcal{G}_{K,\epsilon}.$$

By (8.44), if n is large enough, for each pair (i,j) with $1 \le i < j \le L$ such z_i and z_j can be found so that $z_i \in \Delta_i$ and $z_j \in \Delta_j$. But the Δ_j's were picked to make (8.45) true and so via these pairs of sites we have

$$\left|\chi(y,\omega) - \chi(x,\omega)\right| \le K + \epsilon L + 2\epsilon n \tag{8.47}$$

for every $x,y \in \Delta_1 \cup \cdots \cup \Delta_L$.

From (8.45) and (8.47), we can conclude that for all $r,s \in \Lambda$,

$$\left|\chi(r,\omega) - \chi(s,\omega)\right| \le 3K + \epsilon L + 4\epsilon n < 5\epsilon n, \tag{8.48}$$

provided that $\epsilon n > 3K + \epsilon L$. If $\varrho_{\nu,\epsilon}$ denotes the right-hand side of (8.43) before taking $\epsilon \downarrow 0$, the bounds (8.46) and (8.48) and $\Lambda \cap \Lambda_n^1 \ne \emptyset$ yield

$$\varrho_{\nu+1,5\epsilon}(\omega) \le \delta + L\epsilon,$$

for $\mathbb{P}_\alpha$-a.e. ω. But the left-hand side of this inequality increases as $\epsilon \downarrow 0$ while the right-hand side decreases. Thus, taking $\epsilon \downarrow 0$ and $\delta \downarrow 0$ proves that $\rho_{\nu+1} = 0$ holds for $\mathbb{P}_\alpha$-a.e. ω.   □

Proof of Proposition 8.4.5(2). First, by Lemma 8.7.3, we know that $\varrho_1(\omega) = 0$ for $\mathbb{P}_\alpha$-a.e. ω. Using induction, Lemma 8.7.4 then gives $\varrho_d(\omega) = 0$ for $\mathbb{P}_\alpha$-a.e. ω.

Let $\omega \in \{0 \in C_{\infty,\alpha}\}$. By Lemma 8.7.4, for each $\epsilon > 0$ there exists $n_0 = n_0(\omega)$ which is a.s. finite such that for all $n \geq n_0(\omega)$, we have $|\chi(x,\omega)| \leq \epsilon n$ for all $x \in \Lambda_n^1 \cap C_{\infty,\alpha}(\omega)$. Using this to estimate away the infimum in (8.43), $\varrho_d = 0$ now gives the desired result for all $\epsilon > 0$.   □

8.7.2 Case 2

Recall that for Case 2, we only need to work when $\alpha = 0$, so $C_{\infty,\alpha} = C_\infty = \mathbb{Z}^d$ in this case. We follow [28, Sect. 5]. Let $Q_{x,y}^{(n)} = q_n^\omega(x,y)$. Since the base measure is $\nu_x \equiv 1$, we have $Q_{x,y}^{(n)} = \sum_z Q_{x,z}^{(n-1)} Q_{z,y}^{(1)}$. We first give a preliminary lemma.

Lemma 8.7.5. *For $G \in \overline{L}^2$, we have*

$$\mathbb{E}[\sum_y Q_{0,y}^{(n)} G(\cdot,y))^2] \leq n \|G\|_{\overline{L}^2}^2. \tag{8.49}$$

Proof. Let a_n^2 be the left hand side of (8.49). Then, by the cocycle property, we have

$$a_n^2 = \mathbb{E}[\sum_x \sum_y Q_{0,x}^{(n-1)} Q_{x,y}(G(T_x\cdot, y - x) + G(\cdot,x))^2]$$

$$= \mathbb{E}[\sum_x \sum_y Q_{0,x}^{(n-1)} Q_{x,y}\{G(T_x\cdot, y - x)^2 + 2G(T_x\cdot, y - x)G(\cdot,x) + G(\cdot,x)^2\}]$$

$$=: I + II + III.$$

Then,

$$I = \mathbb{E}[\sum_x \sum_y Q_{-x,0}^{(n-1)}(T_x\cdot)Q_{0,y-x}(T_x\cdot)G(T_x\cdot, y - x)^2]$$

$$= \mathbb{E}[\sum_x \sum_z Q_{-x,0}^{(n-1)}(\cdot)Q_{0,z}(\cdot)G(\cdot,z)^2] = \mathbb{E}[\sum_z Q_{0,z}(\cdot)G(\cdot,z)^2] = \|G\|_{\overline{L}^2}^2,$$

$$III = \mathbb{E}[\sum_x \sum_y Q_{0,x}^{(n-1)} Q_{x,y} G(\cdot,x)^2] = \mathbb{E}[\sum_x Q_{0,x}^{(n-1)} G(\cdot,x)^2] = a_{n-1}^2,$$

$$II = 2\mathbb{E}[\sum_x \sum_z Q_{0,x}^{(n-1)}(\cdot)Q_{0,z}(T_x\cdot)G(T_x\cdot, z)G(\cdot,x)] \leq 2I^{1/2}III^{1/2}$$

$$= 2a_{n-1}\|G\|_{\overline{L}^2}.$$

Thus $a_n \leq a_{n-1} + \|G\|_{\overline{L}^2} \leq \cdots \leq n\|G\|_{\overline{L}^2}$.   □

The next lemma proves Proposition 8.4.5 (1). We use the heat kernel lower bound in the proof.

Lemma 8.7.6. *Let $G \in \overline{L}^2$.*

(i) For $1 \leq p < 2$, there exists $c_p > 0$ such that

$$\mathbb{E}[|G(\cdot, x)|^p]^{1/p} \leq c_p |x| \cdot \|G\|_{\overline{L}^2}, \qquad \forall x \in \mathbb{Z}^d. \tag{8.50}$$

(ii) It follows that $\lim_{n \to \infty} \max_{|x| \leq n} n^{-d-3}|G(\omega, x)| = 0$, $\mathbb{P}$-a.e. ω.

Proof. Using the cocycle property and the triangle inequality iteratively, for $x = (x_1, \cdots, x_d)$, we have

$$\mathbb{E}[|G(\cdot, x)|^p]^{1/p} \leq \sum_{i=1}^{d} |x_i| \mathbb{E}[|G(\cdot, \mathbf{e}_i)|^p]^{1/p},$$

so it is enough to prove $\mathbb{E}[|G(\cdot, \mathbf{e}_i)|^p] \leq c_p \|G\|_{\overline{L}^2}^p$ for $1 \leq i \leq d$. By Theorem 8.1.3, there exists an integer valued random variable W_i with $W_i \geq 1$ such that $\mathbb{P}(W_i = n) \leq c_1 \exp(-c_2 n^\eta)$ for some $\eta > 0$ and $q_t^\omega(0, \mathbf{e}_i) \geq c_3 t^{-d/2}$ for $t \geq W_i$. Set $\xi_{n,i} = q_n^\omega(0, \mathbf{e}_i)$. Then

$$\mathbb{E}[|G(\cdot, \mathbf{e}_i)|^p] = \sum_{n=1}^{\infty} \mathbb{E}[|G(\cdot, \mathbf{e}_i)|^p 1_{\{W_i = n\}}].$$

Let $\alpha = 2/p$ and $\alpha' = 2/(2 - p)$ be its conjugate. Then, using the heat kernel bound and the Hölder inequality,

$$\begin{aligned}
\mathbb{E}[|G(\cdot, \mathbf{e}_i)|^p 1_{\{W_i = n\}}] &= \mathbb{E}[\xi_n^{1/\alpha}|G(\cdot, \mathbf{e}_i)|^p \xi_n^{-1/\alpha} 1_{\{W_i = n\}}] \\
&\leq \mathbb{E}[\xi_n G(\cdot, \mathbf{e}_i)^2]^{1/\alpha} \mathbb{E}[\xi_n^{-\alpha'/\alpha} 1_{\{W_i = n\}}]^{1/\alpha'} \\
&\leq \mathbb{E}[\sum_y Q_{0,y}^{(n)} G(\cdot, y)^2]^{1/\alpha} \left((c_3 n^{-d/2})^{-\alpha'/\alpha} c_1 \exp(-c_2 n^\eta)\right)^{1/\alpha'} \\
&\leq (n\|G\|_{\overline{L}^2}^2)^{1/\alpha} c_4 n^{d/(2\alpha)} \exp(-c_5 n^\eta) \\
&= c_4 n^{(d+2)/(2\alpha)} \exp(-c_5 n^\eta) \|G\|_{\overline{L}^2}^p,
\end{aligned}$$

where we used (8.49) in the last inequality. Summing over $n \geq 1$, we obtain $\mathbb{E}[|G(\cdot, \mathbf{e}_i)|^p] \leq c_p \|G\|_{\overline{L}^2}^p$ for $1 \leq i \leq d$.

(ii) We have

$$\begin{aligned}
\mathbb{P}(\max_{|x| \leq n} |G(\cdot, x)| > \lambda_n) &\leq (2n + 1)^d \max_{|x| \leq n} \mathbb{P}(|G(\cdot, x)| > \lambda_n) \\
&\leq \frac{cn^d}{\lambda_n} \max_{|x| \leq n} \mathbb{E}[|G(\cdot, x)|] \leq \frac{c'n^{d+1}}{\lambda_n} \|G\|_{\overline{L}^2},
\end{aligned}$$

where (8.50) is used in the last inequality. Taking $\lambda_n = n^{d+3}$ and using the Borel-Cantelli, we obtain the desired result.

□

We next prove Proposition 8.4.5 (3-2). One proof is to apply Proposition 8.4.5 (3-1) with $Z(\cdot) \equiv x$. Here is another proof.

Lemma 8.7.7. *For* $G \in L^2_{\text{pot}}$, *it holds that* $\mathbb{E}[G(\cdot, x)] = 0$ *for all* $x \in \mathbb{Z}^d$.

Proof. If $G = \nabla F$ where $F \in L^2$, then $\mathbb{E}[G(\cdot, x)] = \mathbb{E}[F_x - F] = \mathbb{E}F_x - \mathbb{E}F = 0$. In general, there exist $\{F_n\} \subset L^2$ such that $G = \lim_n \nabla F_n$ in $\overline{L}^2$. Since $\mathbb{P}(Q_{0,x} > 0) = 1$ for all $x \in \mathbb{Z}^d$, it follows that $\nabla F_n(\cdot, x)$ converges to $G(\cdot, x)$ in $\mathbb{P}$-probability. By (8.50), for each $p \in [1, 2)$, $\{\nabla F_n(\cdot, x)\}_n$ is bounded in L^p so that $\nabla F_n(\cdot, x)$ converges to $G(\cdot, x)$ in L^1 for all $x \in \mathbb{Z}^d$. Thus $\mathbb{E}[G(\cdot, x)] = \lim_n \mathbb{E}[\nabla F_n(\cdot, x)] = 0$.
□

By this lemma, we have $\mathbb{E}[\chi(\cdot, \mathbf{e}_i)] = 0$ where $\mathbf{e}_i$ is the unit vector for $1 \le i \le d$. Using the cocycle property inductively, we have

$$\chi(\omega, n\mathbf{e}_i) = \sum_{k=1}^{n} \chi(T_{(k-1)\mathbf{e}_i}\omega, \mathbf{e}_i) = \sum_{k=1}^{n} \Psi \circ T_{(k-1)\mathbf{e}_i},$$

where $\Psi(\omega) := \chi(\omega, \mathbf{e}_i)$. By the proof of Lemma 8.7.7, we see that $\Psi \in L^1$. So Birkhoff's ergodic theorem implies

$$\lim_{n \to \infty} \frac{1}{n} \chi(\omega, n\mathbf{e}_i) = 0, \qquad \mathbb{P}\text{-a.e. } \omega. \tag{8.51}$$

Given this, which corresponds to Lemma 8.7.3 for Case 1, the proof of Proposition 8.4.5 (2) for Case 2 is the same (in fact slightly easier since $C_\infty = \mathbb{Z}^d$ in this case) as that of Case 1.

8.8 End of the Proof of Quenched Invariance Principle

In this section, we will complete the proof of Theorem 8.1.2 and Theorem 8.1.4. Throughout this section, we assume that α satisfies (8.7) for Case 1 and $\alpha = 0$ for Case 2.

First, we verify some conditions in Theorem 8.4.3 to prove the sub-linearity of the corrector. The heat kernel estimates and percolation estimates are seriously used here.

Lemma 8.8.1. *(i) The diffusive bounds (8.21), (8.22) hold for* $\mathbb{P}_\alpha$-*a.e.* ω.
(ii) Let $\tau_n = \{t \ge 0 : |\hat{Y}_t - \hat{Y}_0| \ge n\}$. *Then for any* $c_1 > 1$, *there exists* $N_\omega = N_\omega(c_1) > 0$ *which is a.s. finite such that* $|\hat{Y}_{\tau_n} - \hat{Y}_0| \le c_1 n$ *for all* $t > 0$ *and* $n \ge N_\omega$.

Proof. (i) For Case 2, this follows from Theorem 8.1.3. For Case 1, the proof is given in [59, Sect. 6]. (See also [178, Sect. 4]. In fact, in this case we have two-sided Gaussian heat kernel bounds which are similar to the ones for simple random walk on the supercritical percolation clusters.)

(ii) For Case 1, this follows directly from Proposition 8.2.2 and the Borel-Cantelli. For Case 2, using Theorem 8.1.3, we have for any $z \in \mathbb{Z}^d$ with $|z| \leq n$ and $c_1 > 1$,

$$\mathbb{P} \times P_\omega^z(|Y_1| \geq c_1 n) \leq \mathbb{P}(U_z \geq c_1 n) + \mathbb{E}[\sum_{z \in B(0,c_1 n)^c \cap \mathbb{Z}^d} q_1^{\cdot}(z, y) : U_z < c_1 n]$$

$$\leq c_2 \exp(-c_3 n^\eta) + c_4 n^{d-1} \exp(-c_5(c_1 - 1)n \log n),$$

where $c_2, \cdots, c_5 > 0$ do not depend on z. So the result follows by the Borel-Cantelli.

$\square$

Proof of Theorem 8.1.2 and Theorem 8.1.4. By Proposition 8.4.5 and Lemmas 8.8.1, the corrector satisfies the conditions of Theorem 8.4.3. It follows that χ is sub-linear on $\mathcal{C}_{\infty,\alpha}$ (for Case 2, since $\alpha = 0$, χ is sub-linear on $\mathcal{C}_\infty$). However, for Case 1, by (8.8) the diameter of the largest component of $\mathcal{C}_\infty \setminus \mathcal{C}_{\infty,\alpha}$ in a box $[-2n, 2n]$ is less than $C(\omega) \log n$ for some $C(\omega) < \infty$. Using the harmonicity of φ_ω on $\mathcal{C}_\infty$, the optional stopping theorem gives

$$\max_{\substack{x \in \mathcal{C}_\infty \\ |x| \leq n}} |\chi(\omega, x)| \leq \max_{\substack{x \in \mathcal{C}_{\infty,\alpha} \\ |x| \leq n}} |\chi(\omega, x)| + 2C(\omega) \log(2n), \qquad (8.52)$$

hence χ is sub-linear on $\mathcal{C}_\infty$ as well.

Having proved the sub-linearity of χ on $\mathcal{C}_\infty$ for both Case 1 and Case 2, we proceed as in the $d = 2$ proof of [51]. Let $\{Y_t\}_{t \geq 0}$ be the VSRW and $X_n := Y_n$, $n \in \mathbb{N}$ be the discrete time process. Also, let $\varphi_\omega(x) := x + \chi(\omega, x)$ and define $M_n := \varphi_\omega(X_n)$. Fix $\hat{v} \in \mathbb{R}^d$. We will first show that (the piecewise linearization of) $t \mapsto \hat{v} \cdot M_{[tn]}$ scales to Brownian motion. For $K \geq 0$ and for $m \leq n$, let

$$f_K(\omega) := E_\omega^0[(\hat{v} \cdot M_1)^2 1_{\{|\hat{v} \cdot M_1| \geq K\}}], \quad \text{and}$$

$$V_{n,m}(K) := V_{n,m}^{(\omega)}(K) := \frac{1}{n} \sum_{k=0}^{m-1} f_K \circ T_{X_k}(\omega).$$

Since

$$V_{n,m}(\epsilon \sqrt{n}) = \frac{1}{n} \sum_{k=0}^{m-1} E_\omega^0 \left[(\hat{v} \cdot (M_{k+1} - M_k))^2 1_{\{|\hat{v} \cdot (M_{k+1} - M_k)| \geq \epsilon \sqrt{n}\}} \middle| \mathcal{F}_k \right],$$

for $\mathcal{F}_k = \sigma(X_0, X_1, \ldots, X_k)$, in order to apply the Lindeberg-Feller FCLT for martingales (see for example, [95]), we need to verify the following for $\mathbb{P}$-almost every ω:

(i) $V_{n,[tn]}(0) \to Ct$ in P_ω^0-probability for all $t \in [0, 1]$ and some $C \in (0, \infty)$,

(ii) $V_{n,n}(\epsilon \sqrt{n}) \to 0$ in P_ω^0-probability for all $\epsilon > 0$.

By Theorem 8.4.1 (3), $f_K \in \mathbb{L}^1$ for all K. Since $n \mapsto T_{X_n}(\omega)$ (cf. (8.27)) is ergodic, we have

$$V_{n,n}(K) = \frac{1}{n} \sum_{k=0}^{n-1} f_K \circ T_{X_k}(\omega) \xrightarrow[n\to\infty]{} \mathbb{E} f_K, \qquad (8.53)$$

for $\mathbb{P}$-a.e. ω and P_ω^0-a.e. path $\{X_k\}_k$ of the random walk. Taking $K = 0$ in (8.53), condition (i) above follows by scaling out the t-dependence first and working with tn instead of n. On the other hand, by (8.53) and the fact that $f_{\epsilon\sqrt{n}} \le f_K$ for sufficiently large n, we have, $\mathbb{P}$-almost surely,

$$\limsup_{n\to\infty} V_{n,n}(\epsilon \sqrt{n}) \le \mathbb{E} f_K = \mathbb{E}\Big[E_\omega^0\big[(\hat{v} \cdot M_1)^2 1_{\{|\hat{v}\cdot M_1| \ge K\}}\big]\Big] \xrightarrow[K\to\infty]{} 0,$$

where we can use the dominated convergence theorem since $\hat{v} \cdot M_1 \in L^2$. Hence, conditions (i) and (ii) hold (in fact, even with P_ω^0-a.s. limits). We thus conclude that the following continuous random process

$$t \mapsto \frac{1}{\sqrt{n}}\big(\hat{v} \cdot M_{[nt]} + (nt - [nt]) \, \hat{v} \cdot (M_{[nt]+1} - M_{[nt]})\big)$$

converges weakly to Brownian motion with mean zero and covariance $\mathbb{E} f_0 = \mathbb{E} E_\omega^0\big[(\hat{v} \cdot M_1)^2\big]$. This can be written as $\hat{v} \cdot D \hat{v}$ where D is the matrix defined by

$$D_{i,j} := \mathbb{E} E_\omega^0\big((e_i \cdot M_1)(e_j \cdot M_1)\big), \quad \forall i, j \in \{1, \cdots, d\}. \qquad (8.54)$$

Thus, using the Cramér-Wold device and the continuity of the process (see for example, [95]), we obtain that the linear interpolation of $t \mapsto M_{[nt]}/\sqrt{n}$ converges to d-dimensional Brownian motion with covariance matrix D. Since $X_n - M_n = \chi(\omega, X_n)$, in order to prove the convergence of $t \mapsto X_{[nt]}/\sqrt{n}$ to Brownian motion $\mathbb{P}$-a.e. ω, it is enough to show that, for $\mathbb{P}$-a.e. ω,

$$\max_{1 \le k \le n} \frac{|\chi(X_k, \omega)|}{\sqrt{n}} \xrightarrow[n\to\infty]{} 0, \quad \text{in } P_\omega^0\text{-probability.} \qquad (8.55)$$

By the sub-linearity of χ, we know that for each $\epsilon > 0$ there exists $K = K(\omega) < \infty$ such that

$$|\chi(x, \omega)| \le K + \epsilon|x|, \qquad \forall x \in \mathcal{C}_\infty(\omega).$$

Putting $x = X_k$ and substituting $X_k = M_k - \chi(X_k, \omega)$ in the right hand side, we have, if $\epsilon < \frac{1}{2}$,

$$|\chi(X_k, \omega)| \le \frac{K}{1-\epsilon} + \frac{\epsilon}{1-\epsilon}|M_k| \le 2K + 2\epsilon|M_k|.$$

But the above CLT for $\{M_k\}$ and the additivity of $\{M_k\}$ imply that $\max_{k \le n} |M_k|/\sqrt{n}$ converges in law to the maximum of Brownian motion $B(t)$ over $t \in [0, 1]$. Hence, denoting the probability law of Brownian motion by P, we have

$$\limsup_{n \to \infty} P_\omega^0 \Big(\max_{k \le n} |\chi(X_k, \omega)| \ge \delta \sqrt{n} \Big) \le P \Big(\max_{0 \le t \le 1} |B(t)| \ge \frac{\delta}{2\epsilon} \Big).$$

The right-hand side tends to zero as $\epsilon \to 0$ for all $\delta > 0$. Hence we obtain (8.55). We thus conclude that $t \mapsto \hat{B}_n(t)$ converges to d-dimensional Brownian motion with covariance matrix D, where

$$\hat{B}_n(t) := \frac{1}{\sqrt{n}} \big(X_{[tn]} + (tn - [tn])(X_{[tn]+1} - X_{[tn]}) \big), \qquad t \ge 0.$$

Noting that

$$\lim_{n \to \infty} P_\omega^0 \Big(\sup_{0 \le s \le T} |\frac{1}{\sqrt{n}} Y_{sn} - \hat{B}_n(s)| > u \Big) = 0, \qquad \forall u, T > 0,$$

which can be proved using the heat kernel estimates (using Theorem 8.1.3; see [28, Lemma 4.12]) for Case 2, and using the heat kernel estimates for $\hat{Y}$ and the percolation estimate Proposition 8.2.2 for Case 1 (see [178, (3.2)]; note that a VSRW version of [178, (3.2)] that we require can be obtained along the same line of the proof), we see that $t \mapsto Y_{tn}/\sqrt{t}$ converges to the same limit.

By the reflection symmetry of $\mathbb{P}$, we see that D is a diagonal matrix. Further, the rotation symmetry ensures that $D = (\sigma^2/d) I$ where $\sigma^2 := \mathbb{E}_0 E_\omega^0 |M_1|^2$. To see that the limiting process is not degenerate to zero, note that if $\sigma = 0$ then $\chi(\cdot, x) = -x$ holds a.s. for all $x \in \mathbb{Z}^d$. But that is impossible since, as we proved, $x \mapsto \chi(\cdot, x)$ is sub-linear a.s.

Finally we consider the CSRW. For each $x \in C_\infty$, let $\mu_x(\omega) := \sum_y \omega_{xy}$. Let $F(\omega) = \mu_0(\omega)$ and

$$A_t = \int_0^t \mu_{Y_s} ds = \int_0^t F(\omega_s) ds.$$

Then $\tau_t = \inf\{s \ge 0 : A_s > t\}$ is the right continuous inverse of A and $W_t = Y_{\tau_t}$ is the CSRW. Since ω. is ergodic, we have

$$\lim_{t \to \infty} \frac{1}{t} A_t = \mathbb{E} F = 2d \mathbb{E} \mu_e, \qquad \mathbb{P} \times P_\omega^0\text{-a.s.}$$

So if $\mathbb{E}\mu_e < \infty$, then $\tau_t / t \to (2d\mathbb{E}\mu_e)^{-1} =: M > 0$ a.s. Using the heat kernel estimates for Case 2 (i.e. using Theorem 8.1.3; see [28, Theorem 4.11]), and using the heat kernel estimates for $\hat{Y}$ and the percolation estimate Proposition 8.2.2 for Case 1 (i.e. using the VSRW version of [178, (3.2)]), we have

$$\lim_{n \to \infty} P_\omega^0 \Big(\sup_{0 \le s \le T} \Big| \frac{1}{\sqrt{n}} W_{sn} - \frac{1}{\sqrt{n}} Y_{sMn} \Big| > u \Big) = 0, \quad \forall T, u > 0, \;\; \mathbb{P}\text{-a.e. } \omega.$$

Thus, $\frac{1}{\sqrt{n}} W_{sn} = \frac{1}{\sqrt{n}} Y_{sMn} + \frac{1}{\sqrt{n}} W_{sn} - \frac{1}{\sqrt{n}} Y_{sMn}$ converges to $\sigma_C B_t$ where $\{B_t\}$ is Brownian motion and $\sigma_C^2 = M \sigma_V^2 > 0$.

If $\mathbb{E}\mu_e = \infty$, then we have $\tau_t / t \to 0$, so $\frac{1}{\sqrt{n}} W_{sn}$ converges to a degenerate process. $\qquad \square$

Remark 8.8.2. (i) Note that the approach of Mathieu and Piatnitski [179] (also [178]) is different from the above one. They prove sub-linearity of the corrector in the L^2-sense and prove the quenched invariance principle using it. (Here L^2 is with respect to the counting measure on $\mathcal{C}_{\infty,\alpha}$ or on $\mathcal{C}_\infty$. In the above arguments, sub-linearity of the corrector is proved uniformly on $\mathcal{C}_{\infty,\alpha}$, which is much stronger.) Since compactness is crucial in their arguments, they first prove tightness using the heat kernel estimates—recall that the above arguments do not require a separate proof of tightness. Then, using the Poincaré inequality and the so-called two-scale convergence, weak L^2 convergence of the corrector is proved. This together with the Poincaré inequality again, they prove strong L^2 convergence of the corrector (so the L^2-sub-linearity of the corrector). In a sense, they deduce weaker estimates of the corrector (which is however enough for the quenched FCLT) from weaker assumptions.

(ii) The idea that tightness can be proved using heat kernel estimates is already in the paper of Sidoravicius and Sznitman [202]. They also observe that one needs to prove a.s. sub-linearity of the corrector in some sense to get the quenched FCLT.

8.9   General Case (Case 3): $0 \le \mu_e < \infty$

In a recent paper [10], the quenched invariance principle (Theorem 8.1.4) is proved for the general case $0 \le \mu_e < \infty$ with $\mathbb{P}(\mu_e > 0) > p_c(d)$, $d \ge 2$. Here we briefly explain how they establish the theorem.

The basic strategy is in fact similar to that of Case 1. For $K > 1$, let

$$\mathcal{O}_K = E_d \setminus \{e \in E_d : e \cap e' \ne \emptyset \text{ for some } e' \text{ with } \mu_{e'} \in [0, K^{-1}] \cup [K, \infty)\}.$$

By choosing K large enough, one can show that $\mathcal{O}_K$ stochastically dominates a set of bonds in some supercritical percolation [10, Proposition 2.2], so we have a unique infinite cluster of the graph $(\mathbb{Z}^d, \mathcal{O}_K)$ which we denote by $\mathcal{C}_{\infty,K}$. (This corresponds to $\mathcal{C}_{\infty,\alpha}$ for Case 1. The reason we remove neighbors of the "bad bonds" is to keep $\mu_x^{\mathcal{C}_{\infty,K}}$ to be (uniformly) bounded from above and below.) Then, percolation estimates similar to those in Sect. 8.2 hold (see [10, Sect. 2], especially Lemmas 2.3, 2.5 and 2.7).

Let Z be a trace of the original Markov chain Y on $\mathcal{C}_{\infty,K}$, which corresponds to $\hat{Y}$ for Case 1. Then, because "bad bonds" are removed, thanks to the percolation estimates one can obtain detailed heat kernel estimates similar to Theorem 8.1.3 [10, Sect. 4], and the corrector business goes through as in Case 1. As a consequence, one can establish the quenched invariance principle for the trace process Z.

In order to establish the quenched invariance principle for Y, we just compare Y and Z. Since Z is a trace process of Y on $\mathcal{C}_{\infty,K}$, it holds that

$$Z_t = Y_{A_t^{-1}}, \ t \geq 0 \quad \text{where } A_t^{-1} := \inf\{s : A_s > t\}, \ A_s = \int_0^s 1_{\{Y_s \in \mathcal{C}_{\infty,K}\}} ds.$$

By the ergodic theorem, we have

$$\lim_{t \to \infty} \frac{A_t}{t} = \mathbb{P}(0 \in \mathcal{C}_{\infty,K}) =: C_* > 0, \quad \mathbb{P} \times P_\omega^0\text{-a.s.} \tag{8.56}$$

Now write

$$\varepsilon Y_{t/\varepsilon^2} = \varepsilon\left(Y_{t/\varepsilon^2} - Z_{A_{t/\varepsilon^2}}\right) + \varepsilon\left(Z_{A_{t/\varepsilon^2}} - \varepsilon Z_{C_*t/\varepsilon^2}\right) + \varepsilon Z_{C_*t/\varepsilon^2} =: I + II + \varepsilon Z_{C_*t/\varepsilon^2}.$$

Using (8.56) and tightness of Z (which can be deduced using the heat kernel estimates etc.), one can easily prove that II converges to 0 in P_ω^0-probability as $\varepsilon \to 0$, $\mathbb{P}$-a.s. Note that $|Y_s - Z_{A_s}|$ is 0 if $Y_s \in \mathcal{C}_{\infty,K}$ and otherwise it is less than the diameter of the hole containing Y_s. Using the percolation estimate like (8.8) and tightness of Y, it can be proved that I converges to 0 in P_ω^0-probability as $\varepsilon \to 0$, $\mathbb{P}$-a.s. (see [10, Lemma 6.2]). This establishes the convergence of one-dimensional distribution and one can obtain the convergence of finite dimensional distribution similarly. Together with the tightness, the quenched invariance principle for Y is established. As we see, $\sigma_Y^2 = C_* \sigma_Z^2$, where $\sigma_Z^2 I$ is the covariance matrix for the limiting Brownian motion of $\varepsilon Z_{t/\varepsilon^2}$.

Remark 8.9.1. Note that in Case 1, the quenched invariance principle was obtained directly (without proving that of Z) by comparing the correctors as in (8.52). In order to make similar argument though, one needs to define a corrector in a very smart way like in Case 2 (recall that when $\mathbb{E}\mu_e = \infty$, we cannot define correctors for $\{Y_t\}_{t \geq 0}$ in a usual manner).

8.10   Recent Results on the Random Conductance Model

The random conductance model and related models are very actively and exten-
sively studied recently. In this section, we will overview recent results in this area.
Note that they are far from complete.

General Stationary Ergodic Media. It is natural to generalize the quenched
invariance principle in the setting of stationary ergodic media. For Case 0 ($c^{-1} \le$
$\mu_e \le c$), it holds for any stationary ergodic media (see [28, Remark 6.3]). On the
other hand, for Case 2 ($c \le \mu_e < \infty$), there exist stationary ergodic media with
$\mathbb{E}[\mu_e] = \infty$ such that the VSRW can explode in finite time (see for example, [28,
Remark 6.6]).

For $d = 1$ it is well-known (see [55, Exercise 3.12]) that under the assumption

$$\mathbb{E}[\mu_e] < \infty \quad \text{and} \quad \mathbb{E}[\mu_e^{-1}] < \infty, \tag{8.57}$$

the quenched invariance principle holds. In [55, Theorem 4.7], the same result is
established for $d = 2$. In [24], the following is proved; let $p < 1$, then there exists
a stationary ergodic media with

$$\mathbb{E}[\mu_e^p] < \infty \quad \text{and} \quad \mathbb{E}[\mu_e^{-p}] < \infty$$

such that the weak FCLT holds, but the quenched FCLT does not hold for VSRW.
(They give such an example for $d = 2$ and remark it can be generalized for $d \ge 3$
as well.) So one sees that $p = 1$ is critical for the quenched FCLT at least for
$d = 2$. A natural question is whether (8.57) is sufficient for the quenched invariance
principle for $d \ge 3$ or not. In [11], they show the following: let $p, q \in [1, \infty]$ with
$1/p + 1/q < 2/d$, and let $\{\mu_e\}_{e \in E_d}$ be a stationary ergodic media such that

$$\mathbb{P}(0 < \mu_e < \infty) = 1, \quad \mathbb{E}[\mu_e^p] < \infty \quad \text{and} \quad \mathbb{E}[\mu_e^{-q}] < \infty,$$

then the quenched FCLT holds. Their proof of sub-linearity of the corrector uses
Moser's iteration instead of heat kernel estimates. See also a very recent paper [191]
that establishes isoperimetric inequalities and the quenched invariance principle for
a general class of percolation models with long-range correlations.

Stochastic Homogenization. The aim of stochastic homogenization is to relate
stationary ergodic field of coefficients with a deterministic matrix of effective
coefficients. In 1970s, this was firstly done for elliptic equations with random
coefficients—Papanicolaou and Varadhan [189] is one of the first papers in this
direction. If the corrector exists, then it minimizes some Dirichlet energy and
the homogenized coefficients (elements of the covariance matrix for the limiting
Brownian motion) can be expressed using the corrector (as we discussed in (8.54)).
Recently, Gloria, Otto and their collaborators expand the theory of stochastic
homogenization in a discrete setting and introduce quantitative methods that yield
optimal estimates both on the corrector and on approximations of the homogenized
coefficients [113–115]. In [113], which is a continuation and an extension of

[114, 115], they obtain an optimal $\mathbb{L}^p$-decay of the semigroup (in time) for $\hat{L}$ in (8.28) (i.e. semigroup associated with the process "environment seen from the particle") under the assumption of a Poincaré inequality. In particular, for the i.i.d. uniformly elliptic media (i.e. Case 0), they prove existence and $\mathbb{L}^p$-integrability of the corrector for all $1 \leq p < \infty$ as a stationary process in $d > 2$, and also $\mathbb{L}^p$-integrability of the gradient of the corrector. One of the key ingredients of the proof is a new estimate on the gradient of the variable-coefficient heat kernel. In [177], for the uniform elliptic (i.e. Case 0) stationary ergodic media that satisfies a logarithmic Sobolev inequality, they obtain annealed $\mathbb{L}^p$-estimates of the first and second derivative of the Green's function, and applied them to the quantitative estimates on homogenization. For the proof, they use the corresponding $\mathbb{L}^1$-estimates for the second derivative and $\mathbb{L}^2$-estimates for the first derivative for heat kernels that were proved in [90].

These estimates are fundamental and relate to some of the topics discussed below. We expect that these results (and their extended versions) will derive further developments of the RCM in a near future.

Percolation on Half/Quarter Planes. The corrector's method used in previous sections relies on the fact that the environment is stationary and ergodic with respect to the translation on $\mathbb{Z}^d$, so it does not work for half/quarter planes, for example.

In [72] it is proved that the quenched invariance principle holds for random walk on the supercritical percolation cluster on $\mathbb{L} := \{(x_1, \cdots, x_d) \in \mathbb{Z}^d : x_{j_1}, \cdots, x_{j_l} \geq 0\}$ for some $1 \leq j_1 < \cdots < j_l \leq d, l \leq d$. The ideas of the proof are twofold. One is to make a full use of the heat kernel estimates in order to guarantee a subsequential convergence of the scaled processes. (In the previous work, only upper bound of Theorem 8.1.3 was used.) The other is to use the information of the whole space random walk (especially its quenched invariance principle), and to use methods of Dirichlet forms to analyze the behavior around the boundaries. The methods given in the paper are applicable for other models as well.

Speed of Convergence. In [184], it is proved that for Case 2 ($c \leq \mu_e < \infty$) the environment seen from the particle converges to equilibrium polynomially fast in the variance sense. (Note that depending on the depth of the results, additional assumption is given to the environment such as bounded from above, the i.i.d. property.) The results imply some estimate on the speed of convergence of the mean square displacement for the walk towards its limit.

Let $\{X_t\}_t$ be the VSRW. For the i.i.d. uniformly elliptic media (i.e. Case 0), the following Berry-Esseen estimate is established in [185]:

$$\sup_{x \in \mathbb{R}} |\bar{\mathbb{P}}(\frac{\xi \cdot X_t}{\sigma_V \sqrt{t}} \leq x) - \Phi(x)| = \begin{cases} O(t^{-1/10}) & \text{if } d = 1 \\ O(t^{-1/10}(\log t)^q) & \text{if } d = 2 \\ O(t^{-1/5} \log t) & \text{if } d = 3 \\ O(t^{-1/5}) & \text{if } d \geq 4 \end{cases}$$

for some $q \geq 0$ where $\bar{\mathbb{P}}$ is the annealed (averaged) measure and

$$\Phi(x) = (2\pi)^{-1/2} \int_{-\infty}^{x} \exp(-u^2/2) du.$$

In [112], quantitative version of the Kipnis-Varadhan theorem is proved for the i.i.d. uniformly elliptic media (i.e. Case 0), which gives a Monte-Carlo method to approximate effective coefficients for which the convergence rate is dimension-independent. In the work of [112, 185], they use $\mathbb{L}^2$-error estimates given in [114, 115] that are mentioned above.

Growth of Harmonic Functions and Uniqueness of the Corrector. In [47], they develop a quantitative, annealed version of the entropy argument to study the growth of harmonic functions for Markov chains on various random media. In particular, they prove that for the ∞-cluster of supercritical percolation on $\mathbb{Z}^d$, the space of harmonic functions growing at most linearly is $(d + 1)$-dimensional almost surely. Note that the projections of $x + \chi(x)$ (where χ is the corrector) on each coordinate provide d linearly independent harmonic functions of linear growth. So, together with the constant function, the space of (at most) linear growth harmonic functions is at least $(d + 1)$-dimensional. Hence, their result implies that there are no non-constant sub-linear harmonic functions, and implies uniqueness of the corrector (cf. Remark 8.4.2). The entropy argument they develop also provides annealed upper bounds on the space derivative of the heat kernel for various random media (cf. [90, 113, 114]).

Large Deviation. In [165], quenched large deviation principles are proved for random walk on supercritical percolation clusters. It relies on the method developed in the study of quenched large deviation principles for random walk in random environment (see [107] for the corresponding results under some weak elliptic condition and for detailed references). To cope with the lack of ellipticity, [165] uses an estimate on the graph distance due to Antal and Pisztora [15], which was discussed briefly in Proposition 8.2.1 (ii).

In [158], an annealed large deviation principle is established for the normalized local times for VSRW among i.i.d. random conductances in a finite domain with $\log \mathbb{P}(\mu_e \leq \varepsilon) \sim -D\varepsilon^{-\eta}$ as $\varepsilon \to 0$.

Cover Times and Maximum of GFF. Let $\mathcal{C}(n)$ be the largest open cluster for supercritical bond percolation in $[-n, n]^d \cap \mathbb{Z}^d$. In [1], the following is proved; for $d \geq 2$ and $p \in (p_c(d), 1)$, there exists $c_1, c_2 > 0$ such that

$$c_1 n^d (\log n)^2 \leq t_{cov}(\mathcal{C}(n)) \leq c_2 n^d (\log n)^2, \quad \mathbb{P}_p\text{-a.s. for large } n \in \mathbb{N}, \quad (8.58)$$

where $t_{cov}(\mathcal{C}(n))$ is the cover time defined in (7.8). Note that the order of the cover time for d-dimensional torus with size n is $n^d \log n$ for $d \geq 3$ and $n^d (\log n)^2$ for $d = 2$ (see for example, [170, Sect. 11.3.2]). This gives a negative answer to the folk conjecture that most important properties of simple random walks survive percolation (see for example, [190] on this conjecture). Using (7.9), one can also see that the order of the expected maximum of the Gaussian free field for the supercritical percolation is different from that for a box in $d \geq 3$. The key estimate in establishing (8.58) is the following point-to-point resistance estimate:

$$c_1 \log n \leq \max_{x,y \in \mathcal{C}(n)} R_{\text{eff}}^{\mathcal{C}(n)}(x, y) \leq c_2 \log n, \quad \mathbb{P}_p\text{-a.s. for large } n \in \mathbb{N}.$$

A heuristic reason why the order is $\log n$ for $d \geq 3$ is that $\mathcal{C}(n)$ contains many one-dimensional "beards" with size $c \log n$. (Note that for $d = 2$, the resistance order for ordinary random walk is already of order $\log n$, so the effect of the beards cannot be seen.) We note that a similar quenched estimate for $R_{\text{eff}}^{\mathcal{C}_\infty}(x, \partial B_{\mathcal{C}_\infty}(x, n))$ is established in [62], where $\partial B_{\mathcal{C}_\infty}(x, n)$ is the set of boundaries for the ball with respect to the graph distance.

Asymptotic Shape of the Isoperimetric Set. In [57], they discuss geometric properties of isoperimetric sets (i.e. sets with minimal boundary for a given volume) for the infinite cluster of supercritical percolation on $\mathbb{Z}^2$. For a weighted graph (X, μ) with $0 \in X$ and conductance 1 on each bond, define the the isoperimetric profile of X anchored at 0 as

$$\Phi_{X,0}(r) := \inf\left\{ \frac{|\partial \Omega|}{|\Omega|} : 0 \in \Omega \subset X, \text{ restriction of the graph to } \Omega \text{ is connected,} \right.$$

$$\left. 0 < |\Omega| \leq r \right\}.$$

(We may use $\mu(\Omega)$ instead of $|\Omega|$; if (X, μ) has controlled weights, the two values differ up to constant times.) Then, for $d = 2, p > p_c(2) = 1/2$, there exists a constant $\varphi_p \in (0, \infty)$ such that

$$\lim_{n \to \infty} n^{1/2} \Phi_{\mathcal{C}_\infty,0}(n) = \varphi_p \cdot \mathbb{P}(0 \in \mathcal{C}_\infty)^{-1/2}, \quad \mathbb{P}\text{-a.s.}$$

This φ_p can be represented as the isoperimetric constant for a specific isoperimetric problem on $\mathbb{R}^2$, and the asymptotic shape of the isoperimetric set can be characterized by using "Wulff construction" on $\mathbb{R}^2$ (with respect to a particular norm that depends on p). See [57] for details and for references of the related work.

CLT for Effective Conductance. For the i.i.d. uniformly elliptic media (i.e. Case 0), let

$$C_{\text{eff}}^{(N)}(t) := \inf\{\mathcal{E}(f, f) : f(x) = t \cdot x, \, \forall x \in \partial([0, N)^d \cap \mathbb{Z}^d)\}$$

be the effective conductance where $t \in \mathbb{R}^d$ and ∂A is as in (2.6). Because of the choice of the linear boundary condition, subadditivity arguments gives the existence of $\lim_{N \to \infty} N^{-d} C_{\text{eff}}^{(N)}(t)$, $\mathbb{P}$-a.s. In [60], the following is proved: for each $t \in \mathbb{R}^d$, there exists $\sigma_t \in (0, \infty)$ such that

$$\frac{C_{\text{eff}}^{(N)}(t) - \mathbb{E}C_{\text{eff}}^{(N)}(t)}{N^{d/2}} \xrightarrow{\text{law}} \mathcal{N}(0, \sigma_t^2).$$

The proof yields

$$\lim_{N \to \infty} N^{-d} \text{Var}\, (C_{\text{eff}}^{(N)}(t)) = \sigma_t^2.$$

One of the key ingredients for the proof is Meyer's inequality that gives a $\mathbb{L}^p$-estimate for the gradient of solutions of second order elliptic divergence equations (in this case $\mathbb{L}^p$-estimate of $\nabla(\chi + I)$ where χ is the corrector). Note that a torus version of such estimate appeared in [68, Theorem 4.1]. Application of Meyer's inequality in stochastic homogenization (to obtain optimal upper bounds on the variance) goes back to A. Naddaf and T. Spencer (Estimates on the variance of some homogenization problems, 1998, Unpublished manuscript). References [113, 114] discussed above is an important recent work in this direction. The work in [185] is also deeply related.

Biased Random Walks in Random Media. Biased random walks in random media are a natural setting to witness trapping phenomena. For the case of supercritical Galton-Watson trees with leaves, we put conductance β^n where $\beta > 1$ for each bond whose vertex closer to the origin has graph distance n to the origin. This way, the corresponding random walk is β-biased outward the origin. In this case (see [42, 174]) or in the case of supercritical percolation cluster on $\mathbb{Z}^d$ (see [54, 105, 208]), for example, it has been observed that for suitably strong biases dead-ends found in the environment can create a sub-ballistic regime, namely $\lim_{n \to \infty} d(0, X_n)/n = 0$, $\mathbb{P}$-a.s., that is characteristic of trapping. (See also [192] which establishes the quenched invariance principle for directed random walks in space-time random environments.) In the case of randomly biased random walks on a Galton-Watson tree with leaves, it was shown in [132] that precisely the same limiting behavior as the one-dimensional models of [97, 215] occurs. Moreover, there is evidence presented in [105] that suggests the biased random walk on a supercritical percolation cluster also has the same limiting behavior. For IIC of critical Galton-Watson trees, [84] shows that the biased random walk belongs to the same universality class as certain one-dimensional trapping models with slowly-varying tails by establishing functional limit theorems involving an extremal process.

Random Walk on Balanced Random Environment. Let $UE := \{\pm\mathbf{e}_i : i = 1, \cdots, d\}$ be a set of unit vectors in $\mathbb{Z}^d$. Let $\mathcal{M}$ be the space of probability measures on UE and let $\Omega = \mathcal{M}^{\mathbb{Z}^d}$. So, $\omega \in \Omega$ satisfies $\sum_{e \in UE} \omega(z, e) = 1$. Assume the distribution of the environment on Ω is i.i.d., i.e. $\mathbb{P} = \nu^{\mathbb{Z}^d}$ for some $\nu \in \mathcal{M}$. Let $\{X_n\}_n$ be the random walk (Markov chain) on ω, namely

$$P_\omega(X_{n+1} = z + e | X_n = z) = \omega(z, e) \geq 0 \quad \text{for } z \in \mathbb{Z}^d, e \in UE.$$

We say the random walk is balanced and genuinely d-dimensional if

$$P(\omega(z, e) = \omega(z, -e)) = 1 \text{ and } P(\omega(z, e) > 0) > 0, \ \forall z \in \mathbb{Z}^d, e \in UE. \quad (8.59)$$

In [53], it is proved that under (8.59) the quenched invariance principle holds with a deterministic non-degenerate diagonal covariance matrix. Concerning the balanced random walk, in [167] the quenched invariance principle was proved for uniform elliptic (i.e. $\omega(z, e) > \varepsilon_0$ for all $z \in \mathbb{Z}^d$ and $e \in UE$, where ε_0 is some deterministic positive constant) and general ergodic media. In [127] it was proved for elliptic (i.e. $\omega(z, e) > 0$ for all $z \in \mathbb{Z}^d$ and $e \in UE$) and either i.i.d. media or ergodic media with some moment condition.

Einstein Relation. Given a simple random walk X in random media and a unit vector ℓ, consider a biased random walk X^λ i.e. adding a local drift to X in direction ℓ with "intensity" $\lambda > 0$. Assume that $\varepsilon X_{t/\varepsilon^2}$ converges to a Brownian motion with covariance matrix Σ and assume that $\lim_{t\to\infty} E[X_t^\lambda]/t =: v_\lambda$ (effective velocity) exists. Then the following relation is called the Einstein relation:

$$\lim_{\lambda \to 0} v_\lambda/\lambda = \Sigma\ell. \tag{8.60}$$

The name comes from a result in the celebrated paper of Einstein [96]. In the paper, he established a linear relation between the mobility and the diffusivity of a Brownian particle, where the mobility is defined as a derivative of the effective velocity at 0; (8.60) is a natural extension of the relation. A heuristic derivation of this relation can be found, for example in [205, Sect. 8.8].

In [168], the Einstein relation is proved under a general setup in a weak sense (i.e. allowing the velocity to be rescaled with time). Recently, rigorous proofs of the Einstein relation are given for the following random media:

- Tagged particle in symmetric simple exclusions in $d \geq 3$ [171].
- Bond diffusion in $\mathbb{Z}^d$ for special environment distributions [156].
- Mixing dynamical random environments with spectral gap [155].
- Elliptic diffusions in an ergodic random environment with bounded potential and short-range correlations [109].
- Biased random walks on supercritical Galton-Watson trees [43].
- Random walks in a balanced uniformly elliptic i.i.d. environment [126].

There are many on-going work on this topic as well.

References

1. Y. Abe, Effective resistances for supercritical percolation clusters in boxes. ArXiv:1306.5580 (2013)
2. L. Addario-Berry, N. Broutin, C. Goldschmidt, The continuum limit of critical random graphs. Probab. Theory Relat. Fields **152**, 367–406 (2012)
3. M. Aizenman, D.J. Barsky, Sharpness of the phase transition in percolation models. Commun. Math. Phys. **108**, 489–526 (1987)
4. M. Aizenman, C.M. Newman, Tree graph inequalities and critical behavior in percolation models. J. Stat. Phys. **36**, 107–143 (1984)
5. D. Aldous, The continuum random tree. III. Ann. Probab. **21**, 248–289 (1993)
6. D. Aldous, Brownian excursions, critical random graphs and the multiplicative coalescent. Ann. Probab. **25**, 812–854 (1997)
7. D. Aldous, J. Fill, *Reversible Markov Chains and Random Walks on Graphs* (Book in preparation), http://www.stat.berkeley.edu/~aldous/RWG/book.html
8. S. Alexander, R. Orbach, Density of states on fractals: "fractons". J. Phys. (Paris) Lett. **43**, L625–L631 (1982)
9. S. Andres, M.T. Barlow, Energy inequalities for cutoff-functions and some applications. J. Reine Angew. Math. (to appear)
10. S. Andres, M.T. Barlow, J.-D. Deuschel, B.M. Hambly, Invariance principle for the random conductance model. Probab. Theory Relat. Fields **156**, 535–580 (2013)
11. S. Andres, J.-D. Deuschel, M. Slowik, Invariance principle for the random conductance model in a degenerate ergodic environment. ArXiv:1306.2521 (2013)
12. O. Angel, Growth and percolation on the uniform infinite planar triangulation. Geom. Funct. Anal. **13**, 935–974 (2003)
13. O. Angel, J. Goodman, F. den Hollander, G. Slade, Invasion percolation on regular trees. Ann. Probab. **36**, 420–466 (2008)
14. O. Angel, O. Schramm, Uniform infinite planar triangulations. Commun. Math. Phys. **241**, 191–213 (2003)
15. P. Antal, A. Pisztora, On the chemical distance for supercritical Bernoulli percolation. Ann. Probab. **24**, 1036–1048 (1996)
16. M.T. Barlow, Diffusions on fractals, in *Ecole d'été de probabilités de Saint-Flour XXV—1995*. Lecture Notes in Mathematics, vol. 1690 (Springer, New York, 1998)
17. M.T. Barlow, Which values of the volume growth and escape time exponent are possible for a graph? Rev. Mat. Iberoam. **20**, 1–31 (2004)
18. M.T. Barlow, Random walks on supercritical percolation clusters. Ann. Probab. **32**, 3024–3084 (2004)

T. Kumagai, *Random Walks on Disordered Media and their Scaling Limits*, Lecture Notes in Mathematics 2101, DOI 10.1007/978-3-319-03152-1,
© Springer International Publishing Switzerland 2014

19. M.T. Barlow, *Random Walks on Graphs*. Lecture Notes for RIMS Lectures in 2005. Available at http://www.math.kyoto-u.ac.jp/~kumagai/rim10.pdf
20. M.T. Barlow, *Random Walks on Graphs* (Cambridge University Press, to appear)
21. M.T. Barlow, R.F. Bass, Stability of parabolic Harnack inequalities. Trans. Amer. Math. Soc. **356**, 1501–1533 (2003)
22. M.T. Barlow, R.F. Bass, Z.-Q. Chen, M. Kassmann, Non-local Dirichlet forms and symmetric jump processes. Trans. Amer. Math. Soc. **361**, 1963–1999 (2009)
23. M.T. Barlow, R.F. Bass, T. Kumagai, Stability of parabolic Harnack inequalities on metric measure spaces. J. Math. Soc. Jpn. **58**, 485–519 (2006)
24. M.T. Barlow, K. Burdzy, Á. Timár, Comparison of quenched and annealed invariance principles for random conductance model. ArXiv:1304.3498 (2013)
25. M.T. Barlow, J. Černý, Convergence to fractional kinetics for random walks associated with unbounded conductances. Probab. Theory Relat. Fields **149**, 639–673 (2011)
26. M.T. Barlow, T. Coulhon, A. Grigor'yan, Manifolds and graphs with slow heat kernel decay. Invent. Math. **144**, 609–649 (2001)
27. M.T. Barlow, T. Coulhon, T. Kumagai, Characterization of sub-Gaussian heat kernel estimates on strongly recurrent graphs. Commun. Pure Appl. Math. **58**, 1642–1677 (2005)
28. M.T. Barlow, J.-D. Deuschel, Invariance principle for the random conductance model with unbounded conductances. Ann. Probab. **38**, 234–276 (2010)
29. M.T. Barlow, J. Ding, A. Nachmias, Y. Peres, The evolution of the cover time. Comb. Probab. Comput. **20**, 331–345 (2011)
30. M.T. Barlow, A. Grigor'yan, T. Kumagai, On the equivalence of parabolic Harnack inequalities and heat kernel estimates. J. Math. Soc. Jpn. **64**, 1091–1146 (2012)
31. M.T. Barlow, B.M. Hambly, Parabolic Harnack inequality and local limit theorem for random walks on percolation clusters. Electron. J. Probab. **14**, 1–27 (2009)
32. M.T. Barlow, A.A. Járai, T. Kumagai, G. Slade, Random walk on the incipient infinite cluster for oriented percolation in high dimensions. Commun. Math. Phys. **278**, 385–431 (2008)
33. M.T. Barlow, T. Kumagai, Random walk on the incipient infinite cluster on trees. Illinois J. Math. **50**, 33–65 (2006) (electronic)
34. M.T. Barlow, R. Masson, Spectral dimension and random walks on the two dimensional uniform spanning tree. Commun. Math. Phys. **305**, 23–57 (2011)
35. M.T. Barlow, Y. Peres, P. Sousi, Collisions of random walks. Ann. Inst. Henri Poincaré Probab. Stat. **48**, 922–946 (2012)
36. M.T. Barlow, X. Zheng, The random conductance model with Cauchy tails. Ann. Appl. Probab. **20**, 869–889 (2010)
37. D.J. Barsky, M. Aizenman, Percolation critical exponents under the triangle condition. Ann. Probab. **19**, 1520–1536 (1991)
38. R.F. Bass, A stability theorem for elliptic Harnack inequalities. J. Eur. Math. Soc. **17**, 856–876 (2013)
39. G. Ben Arous, M. Cabezas, J. Černý, R. Royfman, Randomly trapped random walks. ArXiv:1302.7227 (2013)
40. G. Ben Arous, J. Černý, Bouchaud's model exhibits two aging regimes in dimension one. Ann. Appl. Probab. **15**, 1161–1192 (2005)
41. G. Ben Arous, J. Černý, Scaling limit for trap models on $\mathbb{Z}^d$. Ann. Probab. **35**, 2356–2384 (2007)
42. G. Ben Arous, A. Fribergh, N. Gantert, A. Hammond, Biased random walks on Galton-Watson trees with leaves. Ann. Probab. **40**, 280–338 (2012)
43. G. Ben Arous, Y. Hu, S. Olla, O. Zeitouni, Einstein relation for biased random walk on Galton-Watson trees. Ann. Inst. Henri Poincaré Probab. Stat. **49**, 698–721 (2013)
44. D. Ben-Avraham, S. Havlin, *Diffusion and Reactions in Fractals and Disordered Systems* (Cambridge University Press, Cambridge, 2000)
45. I. Benjamini, Instability of the Liouville property for quasi-isometric graphs and manifolds of polynomial volume growth. J. Theor. Probab. **4**, 631–637 (1991)

46. I. Benjamini, N. Curien, Simple random walk on the uniform infinite planar quadrangulation: subdiffusivity via pioneer points. Geom. Funct. Anal. **23**, 501–531 (2013)

47. I. Benjamini, H. Duminil-Copin, G. Kozma, A. Yadin, Disorder, entropy and harmonic functions. ArXiv:1111.4853 (2011)

48. I. Benjamini, O. Gurel-Gurevich, R. Lyons, Recurrence of random walk traces. Ann. Probab. **35**, 732–738 (2007)

49. I. Benjamini, E. Mossel, On the mixing time of simple random walk on the super critical percolation cluster. Probab. Theory Relat. Fields **125**, 408–420 (2003)

50. I. Benjamini, O. Schramm, Recurrence of distributional limits of finite planar graphs. Electron. J. Probab. **6**(23), 13 pp. (2001)

51. N. Berger, M. Biskup, Quenched invariance principle for simple random walk on percolation clusters. Probab. Theory Relat. Fields **137**, 83–120 (2007)

52. N. Berger, M. Biskup, C.E. Hoffman, G. Kozma, Anomalous heat-kernel decay for random walk among bounded random conductances. Ann. Inst. Henri Poincaré Probab. Stat. **44**, 374–392 (2008)

53. N. Berger, J.-D. Deuschel, A quenched invariance principle for non-elliptic random walk in i.i.d. balanced random environment. Probab. Theory Relat. Fields (to appear)

54. N. Berger, N. Gantert, Y. Peres, The speed of biased random walk on percolation clusters. Probab. Theory Relat. Fields **126**, 221–242 (2003)

55. M. Biskup, Recent progress on the random conductance model. Probab. Surv. **8**, 294–373 (2011)

56. M. Biskup, O. Boukhadra, Subdiffusive heat-kernel decay in four-dimensional i.i.d. random conductance models. J. Lond. Math. Soc. (2) **86**, 455–481 (2012)

57. M. Biskup, O. Louidor, E.B. Procaccia, R. Rosenthal, Isoperimetry in two-dimensional percolation. ArXiv:1211.0745 (2012)

58. M. Biskup, O. Louidor, A. Rozinov, A. Vandenberg-Rodes, Trapping in the random conductance model. J. Stat. Phys. **150**, 66–87 (2013)

59. M. Biskup, T.M. Prescott, Functional CLT for random walk among bounded random conductances. Electron. J. Probab. **12**, 1323–1348 (2007)

60. M. Biskup, M. Salvi, T. Wolff, A central limit theorem for the effective conductance: I. Linear boundary data and small ellipticity contrasts. ArXiv:1210.2371 (2012)

61. M. Biskup, H. Spohn, Scaling limit for a class of gradient fields with non-convex potentials. Ann. Probab. **39**, 224–251 (2011)

62. D. Boivin, C. Rau, Existence of the harmonic measure for random walks on graphs and in random environments. J. Stat. Phys. **150**, 235–263 (2013)

63. B. Bollobás, *Random Graphs*, 2nd edn. (Cambridge University Press, Cambridge, 2001)

64. E. Bolthausen, A.-S. Sznitman, *Ten Lectures on Random Media*. DMV Seminar, vol. 32 (Birkhäuser Verlag, Basel, 2002)

65. O. Boukhadra, Heat-kernel estimates for random walk among random conductances with heavy tail. Stoch. Process. Their Appl. **120**, 182–194 (2010)

66. O. Boukhadra, P. Mathieu, The polynomial lower tail random conductances model. ArXiv:1308.1067 (2013)

67. K. Burdzy, G.F. Lawler, Rigorous exponent inequalities for random walks. J. Phys. A Math. Gen. **23**, L23–L28 (1999)

68. P. Caputo, D. Ioffe, Finite volume approximation of the effective diffusion matrix: the case of independent bond disorder. Ann. Inst. Henri Poincaré Probab. Stat. **39**, 505–525 (2003)

69. E.A. Carlen, S. Kusuoka, D.W. Stroock, Upper bounds for symmetric Markov transition functions. Ann. Inst. Henri Poincaré Probab. Stat. **23**, 245–287 (1987)

70. J. Černý, On two-dimensional random walk among heavy-tailed conductances. Electron. J. Probab. **16**, 293–313 (2011, Paper no. 10)

71. D. Chen, X. Chen, Two random walks on the open cluster of $\mathbb{Z}^2$ meet infinitely often. Sci. China Math. **53**, 1971–1978 (2010)

72. Z.-Q. Chen, D. Croydon, T. Kumagai, Quenched invariance principles for random walks and elliptic diffusions in random media with boundary. ArXiv:1306.0076 (2013)

73. Z.-Q. Chen, P. Kim, T. Kumagai, Discrete approximation of symmetric jump processes on metric measure spaces. Probab. Theory Relat. Fields **155**, 703–749 (2013)

74. Z.-Q. Chen, M. Fukushima, *Symmetric Markov Processes, Time Change, and Boundary Theory* (Princeton University Press, Princeton, 2011)

75. T. Coulhon, Heat kernel and isoperimetry on non-compact Riemannian manifolds, in *Contemporary Mathematics*, vol. 338 (American Mathematical Society, Providence, 2003), pp. 65–99

76. T. Coulhon, Ultracontractivity and Nash type inequalities. J. Funct. Anal. **141**, 510–539 (1996)

77. T. Coulhon, L. Saloff-Coste, Variétés riemanniennes isométriques à l'infini. Rev. Mat. Iberoam. **11**, 687–726 (1995)

78. N. Crawford, A. Sly, Simple random walks on long range percolation clusters I: heat kernel bounds. Probab. Theory Relat. Fields **154**, 753–786 (2012)

79. N. Crawford, A. Sly, Simple random walks on long range percolation clusters II: scaling limits. Ann. Probab. **41**, 445–502 (2013)

80. D.A. Croydon, Heat kernel fluctuations for a resistance form with non-uniform volume growth. Proc. Lond. Math. Soc. (3) **94**, 672–694 (2007)

81. D.A. Croydon, Convergence of simple random walks on random discrete trees to Brownian motion on the continuum random tree. Ann. Inst. Henri Poincaré Probab. Stat. **44**, 987–1019 (2008)

82. D.A. Croydon, Random walk on the range of random walk. J. Stat. Phys. **136**, 349–372 (2009)

83. D.A. Croydon, Scaling limit for the random walk on the largest connected component of the critical random graph. Publ. Res. Inst. Math. Sci. **48**, 279–338 (2012)

84. D. Croydon, A. Fribergh, T. Kumagai, Biased random walk on critical Galton-Watson trees conditioned to survive. Probab. Theory Relat. Fields **157**, 453–507 (2013)

85. D.A. Croydon, B.M. Hambly, Local limit theorems for sequences of simple random walks on graphs. Potential Anal. **29**, 351–389 (2008)

86. D.A. Croydon, B.M. Hambly, T. Kumagai, Convergence of mixing times for sequences of random walks on finite graphs. Electron. J. Probab. **17**(3), 32 pp. (2012)

87. D. Croydon, T. Kumagai, Random walks on Galton-Watson trees with infinite variance offspring distribution conditioned to survive. Electron. J. Probab. **13**, 1419–1441 (2008)

88. M. Damron, J. Hanson, P. Sosoe, Subdiffusivity of random walk on the 2D invasion percolation cluster. Stoch. Process. Their Appl. **123**, 3588–3621 (2013)

89. T. Delmotte, Parabolic Harnack inequality and estimates of Markov chains on graphs. Rev. Mat. Iberoam. **15**, 181–232 (1999)

90. T. Delmotte, J.-D. Deuschel, On estimating the derivatives of symmetric diffusions in stationary random environment, with applications to $\nabla\phi$ interface model. Probab. Theory Relat. Fields **133**, 358–390 (2005)

91. A. De Masi, P.A. Ferrari, S. Goldstein, W.D. Wick, An invariance principle for reversible Markov processes. Applications to random motions in random environments. J. Stat. Phys. **55**, 787–855 (1989)

92. J. Ding, J.R. Lee, Y. Peres, Cover times, blanket times and majorizing measures. Ann. Math. **175**, 1409–1471 (2012)

93. P.G. Doyle, J.L. Snell, *Random Walks and Electric Networks* (Mathematical Association of America, Washington, 1984). ArXiv:math/0001057

94. A. Drewitz, A.F. Ramírez, Selected topics in random walk in random environment. ArXiv:1309.2589 (2013)

95. R. Durrett, *Probability: Theory and Examples*, 4th edn. (Cambridge University Press, Cambridge, 2010)

96. A. Einstein, Die von der molekularkinetischen Theorie der Wärme geforderte Bewegung von in ruhenden Flüssigkeit suspendierten Teilchen. Ann. d. Phys. **17**, 549–560 (1905)

97. N. Enriquez, C. Sabot, O. Zindy, Limit laws for transient random walks in random environment on $\mathbb{Z}$. Ann. Inst. Fourier (Grenoble) **59**, 2469–2508 (2009)

98. P. Erdős, A. Rényi, On the evolution of random graphs. Magyar Tud. Akad. Mat. Kutató Int. Közl. **5**, 17–61 (1960)

99. E.B. Fabes, D.W. Stroock, A new proof of Moser's parabolic Harnack inequality using the old ideas of Nash. Arch. Ration. Mech. Anal. **96**, 327–338 (1986)

100. P.A. Ferrari, R.M. Grisi, P. Groisman, Harmonic deformation of Delaunay triangulations. Stoch. Process. Their Appl. **122**, 2185–2210 (2012)

101. R. Fitzner, Non-backtracking lace expansion, Ph.D. Thesis, The Eindhoven University of Technology, 2013, http://www.win.tue.nl/~rfitzner/NoBLE/index.html

102. R. Fitzner, R. van der Hofstad, Nearest-neighbor percolation function is continuous for $d \geq 15$. (Tentative title) (2013, in preparation)

103. L.R.G. Fontes, M. Isopi, C.M. Newman, Random walks with strongly inhomogeneous rates and singular diffusions: convergence, localization and aging in one dimension. Ann. Probab. **30**, 579–604 (2002)

104. L.R.G. Fontes, P. Mathieu, On symmetric random walks with random conductances on $\mathbb{Z}^d$. Probab. Theory Relat. Fields **134**, 565–602 (2006)

105. A. Fribergh, A. Hammond, Phase transition for the speed of the biased random walk on the supercritical percolation cluster. Commun. Pure Appl. Math. (to appear)

106. I. Fujii, T. Kumagai, Heat kernel estimates on the incipient infinite cluster for critical branching processes, in *Proceedings of German-Japanese Symposium in Kyoto 2006*, RIMS Kôkyûroku Bessatsu, vol. B6, 2008, pp. 85–95

107. R. Fukushima, N. Kubota, Quenched large deviations for multidimensional random walk in random environment with holding times. J. Theor. Probab. (to appear)

108. M. Fukushima, Y. Oshima, M. Takeda, *Dirichlet Forms and Symmetric Markov Processes* (de Gruyter, Berlin, 1994)

109. N. Gantert, P. Mathieu, A. Piatnitski, Einstein relation for reversible diffusions in a random environment. Commun. Pure Appl. Math. **65**, 187–228 (2012)

110. P.G. de Gennes, La percolation: un concept unificateur. La Recherche **7**, 919–927 (1976)

111. G. Giacomin, S. Olla, H. Spohn, Equilibrium fluctuations for $\nabla\phi$ interface model. Ann. Probab. **29**, 1138–1172 (2001)

112. A. Gloria, J.-C. Mourrat, Quantitative version of the Kipnis-Varadhan theorem and Monte Carlo approximation of homogenized coefficients. Ann. Appl. Probab. **23**, 1544–1583 (2013)

113. A. Gloria, S. Neukamm, F. Otto, Quantification of ergodicity in stochastic homogenization: optimal bounds via spectral gap on Glauber dynamics. Preprint 2013

114. A. Gloria, F. Otto, An optimal variance estimate in stochastic homogenization of discrete elliptic equations. Ann. Probab. **39**, 779–856 (2011)

115. A. Gloria, F. Otto, An optimal error estimate in stochastic homogenization of discrete elliptic equations. Ann. Appl. Probab. **22**, 1–28 (2012)

116. A. Grigor'yan, The heat equation on non-compact Riemannian manifolds. Matem. Sbornik. **182**, 55–87 (1991, in Russian) (English transl.) Math. USSR Sb. **72**, 47–77 (1992)

117. A. Grigor'yan, Heat kernel upper bounds on a complete non-compact manifold. Rev. Mat. Iberoam. **10**, 395–452 (1994)

118. A. Grigor'yan, *Analysis on Graphs* (2012). http://www.math.uni-bielefeld.de/~grigor/aglect.pdf

119. A. Grigor'yan, L. Saloff-Coste, Heat kernel on manifolds with ends. Ann. Inst. Fourier (Grenoble) **59**, 1917–1997 (2009)

120. A. Grigor'yan, A. Telcs, Sub-Gaussian estimates of heat kernels on infinite graphs. Duke Math. J. **109**, 452–510 (2001)

121. A. Grigor'yan, A. Telcs, Harnack inequalities and sub-Gaussian estimates for random walks. Math. Annalen **324**, 521–556 (2002)

122. G. Grimmett, *Percolation*, 2nd edn. (Springer, Berlin, 1999)

123. G. Grimmett, P. Hiemer, Directed percolation and random walk, in *In and Out of Equilibrium (Mambucaba, 2000)*. Progress in Probability, vol. 51 (Birkhäuser, Boston, 2002), pp. 273–297

124. M. Gromov, Hyperbolic manifolds, groups and actions, in *Riemann Surfaces and Related Topics: Proceedings of the 1978 Stony Brook Conference*, State University of New York, Stony Brook, 1978. Annals of Mathematics Studies, vol. 97 (Princeton University Press, Princeton, 1981), pp. 183–213

125. A. Guionnet, B. Zegarlinksi, Lectures on logarithmic Sobolev inequalities, in *Séminaire de Probabilités XXXVI*. Lecture Notes in Mathematics, vol. 1801 (Springer, Berlin, 2003), pp. 1–134

126. X. Guo, Einstein relation for random walks in balanced random environment. ArXiv:1212.0255 (2012)

127. X. Guo, O. Zeitouni, Quenched invariance principle for random walks in balanced random environment. Probab. Theory Relat. Fields **152**, 207–230 (2012)

128. O. Gurel-Gurevich, A. Nachmias, Recurrence of planar graph limits. Ann. Math. (2) **177**, 761–781 (2013)

129. P. Hajłasz, P. Koskela, Sobolev met Poincaré. Mem. Amer. Math. Soc. **145**(688), x+101 pp. (2000)

130. B.M. Hambly, T. Kumagai, Heat kernel estimates for symmetric random walks on a class of fractal graphs and stability under rough isometries, in *Fractal Geometry and Applications: A Jubilee of B. Mandelbrot, Proceedings of Symposia in Pure Mathematics*, vol. 72, Part 2 (American Mathematical Society, Providence, 2004), pp. 233–260

131. B.M. Hambly, T. Kumagai, Diffusion on the scaling limit of the critical percolation cluster in the diamond hierarchical lattice. Commun. Math. Phys. **295**, 29–69 (2010)

132. A. Hammond, Stable limit laws for randomly biased walks on supercritical trees. Ann. Probab. **41**, 1694–1766 (2013)

133. T. Hara, Decay of correlations in nearest-neighbour self-avoiding walk, percolation, lattice trees and animals. Ann. Probab. **36**, 530–593 (2008)

134. T. Hara, R. van der Hofstad, G. Slade, Critical two-point functions and the lace expansion for spread-out high-dimensional percolation and related models. Ann. Probab. **31**, 349–408 (2003)

135. M. Heydenreich, R. van der Hofstad, Random graph asymptotics on high-dimensional tori II: volume, diameter and mixing time. Probab. Theory Relat. Fields **149**, 397–415 (2011)

136. M. Heydenreich, R. van der Hofstad, T. Hulshof, Random walk on the high-dimensional IIC. ArXiv:1207.7230 (2012)

137. R. van der Hofstad, F. den Hollander, G. Slade, Construction of the incipient infinite cluster for spread-out oriented percolation above $4 + 1$ dimensions. Commun. Math. Phys. **231**, 435–461 (2002)

138. R. van der Hofstad, A.A. Járai, The incipient infinite cluster for high-dimensional unoriented percolation. J. Stat. Phys. **114**, 625–663 (2004)

139. R. van der Hofstad, G. Slade, Convergence of critical oriented percolation to super-Brownian motion above $4 + 1$ dimensions. Ann. Inst. Henri Poincaré Probab. Stat. **39**, 415–485 (2003)

140. B.D. Hughes, *Random Walks and Random Environments, Volume 2: Random Environments* (Oxford University Press, Oxford, 1996)

141. A.A. Járai, A. Nachmias, Electrical resistance of the low dimensional critical branching random walk. ArXiv:1305.1092 (2013)

142. D. Jerison, The Poincaré inequality for vector fields satisfying Hörmander's condition. Duke Math. J. **53**, 503–523 (1986)

143. V.V. Jikov, S.M. Kozlov, O.A. Oleinik, *Homogenization of Differential Operators and Integral Functionals* (Springer, Berlin, 1994)

144. O.D. Jones, Transition probabilities for the simple random walk on the Sierpinski graph. Stoch. Process. Their Appl. **61**, 45–69 (1996)

145. M. Kanai, Rough isometries, and combinatorial approximations of geometries of non-compact riemannian manifolds. J. Math. Soc. Jpn. **37**, 391–413 (1985)

146. M. Kanai, Analytic inequalities, and rough isometries between non-compact riemannian manifolds, in *Lecture Notes in Mathematics*, vol. 1201 (Springer, Berlin, 1986), pp. 122–137

147. A. Kasue, Convergence of metric graphs and energy forms. Rev. Mat. Iberoam. **26**, 367–448 (2010)

148. H. Kesten, *Percolation Theory for Mathematicians*. Progress in Probability and Statistics, vol. 2 (Birkhäuser, Boston, 1982), iv+423 pp. Available at http://www.math.cornell.edu/~kesten/kesten-book.html

149. H. Kesten, The incipient infinite cluster in two-dimensional percolation. Probab. Theory Relat. Fields **73**, 369–394 (1986)

150. H. Kesten, Subdiffusive behavior of random walk on a random cluster. Ann. Inst. Henri Poincaré Probab. Stat. **22**, 425–487 (1986)

151. J. Kigami, *Analysis on Fractals* (Cambridge University Press, Cambridge, 2001)

152. J. Kigami, Dirichlet forms and associated heat kernels on the Cantor set induced by random walks on trees. Adv. Math. **225**, 2674–2730 (2010)

153. J. Kigami, Resistance forms, quasisymmetric maps and heat kernel estimates. Mem. Amer. Math. Soc. **216**(1015), vi+132 pp. (2012)

154. C. Kipnis, S.R.S. Varadhan, Central limit theorem for additive functionals of reversible Markov processes and applications to simple exclusions. Commun. Math. Phys. **104**, 1–19 (1986)

155. T. Komorowski, S. Olla, On mobility and Einstein relation for tracers in time-mixing random environments. J. Stat. Phys. **118**, 407–435 (2005)

156. T. Komorowski, S. Olla, Einstein relation for random walks in random environments. Stoch. Process. Their Appl. **115**, 1279–1301 (2005)

157. T. Komorowski, C. Landim, S. Olla, *Fluctuations in Markov Processes: Time Symmetry and Martingale Approximation*. Grundlehren der mathematischen Wissenschaften, vol. 345 (Springer, Berlin, 2012)

158. W. König, M. Salvi, T. Wolff, Large deviations for the local times of a random walk among random conductances. Electron. Commun. Probab. **17**, 1–11 (2012, Paper no. 10)

159. S.M. Kozlov, The method of averaging and walks in inhomogeneous environments. Russ. Math. Surv. **40**, 73–145 (1985)

160. G. Kozma, Percolation on a product of two trees. Ann. Probab. **39**, 1864–1895 (2011)

161. G. Kozma, The triangle and the open triangle. Ann. Inst. Henri Poincaré Probab. Stat. **47**, 75–79 (2011)

162. G. Kozma, A. Nachmias, The Alexander-Orbach conjecture holds in high dimensions. Invent. Math. **178**, 635–654 (2009)

163. G. Kozma, A. Nachmias, Arm exponents in high dimensional percolation. J. Amer. Math. Soc. **24**, 375–409 (2011)

164. M. Krikun, Local structure of random quadrangulations. ArXiv:math/0512304 (2005)

165. N. Kubota, Large deviations for simple random walk on supercritical percolation clusters. Kodai Math. J. **35**, 560–575 (2012)

166. T. Kumagai, J. Misumi, Heat kernel estimates for strongly recurrent random walk on random media. J. Theor. Probab. **21**, 910–935 (2008)

167. G.F. Lawler, Weak convergence of a random walk in a random environment. Commun. Math. Phys. **87**, 81–87 (1982)

168. J.L. Lebowitz, H. Rost, The Einstein relation for the displacement of a test particle in a random environment. Stoch. Process. Their Appl. **54**, 183–196 (1994)

169. J.-F. Le Gall, G. Miermont, Scaling limits of random trees and planar maps. Lecture Notes for the Clay Mathematical Institute Summer School in Buzios, 2010. ArXiv:1101.4856

170. D. Levin, Y. Peres, E. Wilmer, *Markov Chains and Mixing Times* (American Mathematical Society, Providence, 2009)

171. M. Loulakis, Einstein relation for a tagged particle in simple exclusion processes. Commun. Math. Phys. **229**, 347–367 (2002)

172. L. Lovász, P. Winkler, Mixing times, in *Microsurveys in Discrete Probability (Princeton, NJ, 1997)*. DIMACS Series in Discrete Mathematics and Theoretical Computer Science, vol. 41 (American Mathematical Society, Providence, 1998), pp. 85–133

173. F. Lust-Piquard, Lower bounds on $\| K^n \|_{1 \to \infty}$ for some contractions K of $L^2(\mu)$, with some applications to Markov operators. Math. Ann. **303**, 699–712 (1995)

174. R. Lyons, R. Pemantle, Y. Peres, Biased random walks on Galton-Watson trees. Probab. Theory Relat. Fields **106**, 249–264 (1996)

175. R. Lyons, Y. Peres, *Probability on Trees and Networks* (Book in preparation), http://mypage.iu.edu/~rdlyons/prbtree/prbtree.html

176. T. Lyons, Instability of the Liouville property for quasi-isometric Riemannian manifolds and reversible Markov chains. J. Differ. Geom. **26**, 33–66 (1987)

177. D. Marahrens, F. Otto, Annealed estimates on the Green's function. ArXiv:1304.4408 (2013)

178. P. Mathieu, Quenched invariance principles for random walks with random conductances. J. Stat. Phys. **130**, 1025–1046 (2008)

179. P. Mathieu, A. Piatnitski, Quenched invariance principles for random walks on percolation clusters. Proc. R. Soc. A **463**, 2287–2307 (2007)

180. P. Mathieu, E. Remy, Isoperimetry and heat kernel decay on percolation clusters. Ann. Probab. **32**, 100–128 (2004)

181. R. Montenegro, P. Tetali, Mathematical aspects of mixing times in Markov chains. Found. Trends Theor. Comput. Sci. **1**, x+121 pp. (2006)

182. B. Morris, Y. Peres, Evolving sets, mixing and heat kernel bounds. Probab. Theory Relat. Fields **133**, 245–266 (2005)

183. J.C. Mourrat, Scaling limit of the random walk among random traps on $\mathbb{Z}^d$. Ann. Inst. Henri Poincaré Probab. Stat. **47**, 813–849 (2011)

184. J.C. Mourrat, Variance decay for functionals of the environment viewed by the particle. Ann. Inst. Henri Poincaré Probab. Stat. **47**, 294–327 (2011)

185. J.C. Mourrat, A quantitative central limit theorem for the random walk among random conductances. Electron. J. Probab. **17**(97), 17 pp. (2012)

186. A. Nachmias, Y. Peres, Critical random graphs: diameter and mixing time. Ann. Probab. **36**, 1267–1286 (2008)

187. J. Nash, Continuity of solutions of parabolic and elliptic equations. Amer. J. Math. **80**, 931–954 (1958)

188. J. Norris, *Markov Chains* (Cambridge University Press, Cambridge, 1998)

189. G.C. Papanicolaou, S.R.S. Varadhan, Boundary value problems with rapidly oscillating random coefficients, in *Random Fields, Vol. I, II (Esztergom, 1979)*. Colloq. Math. Soc. János Bolyai, vol. 27 (North-Holland, Amsterdam, 1981), pp. 835–873

190. G. Pete, A note on percolation on $\mathbb{Z}^d$: isoperimetric profile via exponential cluster repulsion. Electron. Commun. Probab. **13**, 377–392 (2008)

191. E. Procaccia, R. Rosenthal, A. Sapozhnikov, Quenched invariance principle for simple random walk on clusters in correlated percolation models. ArXiv:1310.4764 (2013)

192. F. Rassoul-Agha, T. Seppäläinen, An almost sure invariance principle for random walks in a space-time random environment. Probab. Theory Relat. Fields **133**, 299–314 (2005)

193. C. Rau, Sur le nombre de points visités par une marche aléatoire sur un amas infini de percolation. Bull. Soc. Math. France **135**, 135–169 (2007)

194. L. Saloff-Coste, A note on Poincaré, Sobolev, and Harnack inequalities. Int. Math. Res. Not. IMRN **2**, 27–38 (1992)

195. L. Saloff-Coste, Lectures on finite Markov chains, in *Ecole d'Eté de Probabilités de Saint-Flour XXVI-1996*. Lecture Notes in Mathematics, vol. 1665 (Springer, Berlin, 1997), pp. 301–413

196. L. Saloff-Coste, *Aspects of Sobolev-Type Inequalities*. London Mathematical Society Lecture Notes, vol. 289 (Cambridge University Press, Cambridge, 2002)

197. A. Sapozhnikov, Upper bound on the expected size of intrinsic ball. Electron. Commun. Probab. **15**, 297–298 (2010)

198. R. Schonmann, Multiplicity of phase transitions and mean-field criticality on highly non-amenable graphs. Commun. Math. Phys. **219**, 271–322 (2001)

199. D. Shiraishi, Heat kernel for random walk trace on $\mathbb{Z}^3$ and $\mathbb{Z}^4$. Ann. Inst. Henri Poincaré Probab. Stat. **46**, 1001–1024 (2010)

200. D. Shiraishi, Exact value of the resistance exponent for four dimensional random walk trace. Probab. Theory Relat. Fields **153**, 191–232 (2012)
201. D. Shiraishi, Random walk on non-intersecting two-sided random walk trace is subdiffusive in low dimensions. Trans. Amer. Math. Soc. (to appear)
202. V. Sidoravicius, A.-S. Sznitman, Quenched invariance principles for walks on clusters of percolation or among random conductances. Probab. Theory Relat. Fields **129**, 219–244 (2004)
203. G. Slade, *The Lace Expansion and its Applications. Ecole d'Eté de Probabilités de Saint-Flour XXXIV—2004*. Lecture Notes in Mathematics, vol. 1879 (Springer, Berlin, 2006)
204. P.M. Soardi, *Potential Theory on Infinite Networks*. Lecture Notes in Mathematics, vol. 1590 (Springer, Berlin, 1994)
205. H. Spohn, *Large Scale Dynamics of Interacting Particles* (Springer, Berlin, 1991)
206. R.S. Strichartz, *Differential Equations on Fractals: A Tutorial* (Princeton University Press, Princeton, 2006)
207. K.T. Sturm, Analysis on local Dirichlet spaces—III. The parabolic Harnack inequality. J. Math. Pure Appl. **75**, 273–297 (1996)
208. A-S. Sznitman, On the anisotropic walk on the supercritical percolation cluster. Commun. Math. Phys. **240**, 123–148 (2003)
209. A-S. Sznitman, Topics in random walks in random environment, in *School and Conference on Probability Theory*. ICTP Lecture Notes, vol. XVII (Abdus Salam International Centre for Theoretical Physics, Trieste, 2004), pp. 203–266 (electronic)
210. A. Telcs, *The Art of Random Walks*. Lecture Notes in Mathematics, vol. 1885 (Springer, Berlin, 2006)
211. W. Woess, in *Random Walks on Infinite Graphs and Groups* (Cambridge University Press, Cambridge, 2000)
212. G.M. Zaslavsky, *Hamiltonian Chaos and Fractional Dynamics* (Oxford University Press, Oxford, 2005)
213. O. Zeitouni, Random walks in random environment, in *Ecole d'Eté de Probabilités de Saint-Flour XXXI—2001*. Lecture Notes in Mathematics, vol. 1837 (Springer, Berlin, 2004), pp. 189–312
214. O. Zeitouni, Random walks and diffusions in random environments. J. Phys. A Math. Gen. **39**, R433–R464 (2006)
215. O. Zindy, Scaling limit and aging for directed trap models. Markov Process. Relat. Fields **15**, 31–50 (2009)

Index

T. Kumagai, *Random Walks on Disordered Media and their Scaling Limits*, Lecture Notes 145
in Mathematics 2101, DOI 10.1007/978-3-319-03152-1,
© Springer International Publishing Switzerland 2014

LECTURE NOTES IN MATHEMATICS  Springer

Edited by J.-M. Morel, B. Teissier; P.K. Maini

Editorial Policy (for the publication of monographs)

1. Lecture Notes aim to report new developments in all areas of mathematics and their applications - quickly, informally and at a high level. Mathematical texts analysing new developments in modelling and numerical simulation are welcome.

   Monograph manuscripts should be reasonably self-contained and rounded off. Thus they may, and often will, present not only results of the author but also related work by other people. They may be based on specialised lecture courses. Furthermore, the manuscripts should provide sufficient motivation, examples and applications. This clearly distinguishes Lecture Notes from journal articles or technical reports which normally are very concise. Articles intended for a journal but too long to be accepted by most journals, usually do not have this "lecture notes" character. For similar reasons it is unusual for doctoral theses to be accepted for the Lecture Notes series, though habilitation theses may be appropriate.

2. Manuscripts should be submitted either online at www.editorialmanager.com/lnm to Springer's mathematics editorial in Heidelberg, or to one of the series editors. In general, manuscripts will be sent out to 2 external referees for evaluation. If a decision cannot yet be reached on the basis of the first 2 reports, further referees may be contacted: The author will be informed of this. A final decision to publish can be made only on the basis of the complete manuscript, however a refereeing process leading to a preliminary decision can be based on a pre-final or incomplete manuscript. The strict minimum amount of material that will be considered should include a detailed outline describing the planned contents of each chapter, a bibliography and several sample chapters.

   Authors should be aware that incomplete or insufficiently close to final manuscripts almost always result in longer refereeing times and nevertheless unclear referees' recommendations, making further refereeing of a final draft necessary.

   Authors should also be aware that parallel submission of their manuscript to another publisher while under consideration for LNM will in general lead to immediate rejection.

3. Manuscripts should in general be submitted in English. Final manuscripts should contain at least 100 pages of mathematical text and should always include

   - a table of contents;
   - an informative introduction, with adequate motivation and perhaps some historical remarks: it should be accessible to a reader not intimately familiar with the topic treated;
   - a subject index: as a rule this is genuinely helpful for the reader.

   For evaluation purposes, manuscripts may be submitted in print or electronic form (print form is still preferred by most referees), in the latter case preferably as pdf- or zipped ps-files. Lecture Notes volumes are, as a rule, printed digitally from the authors' files. To ensure best results, authors are asked to use the LaTeX2e style files available from Springer's web-server at:

   ftp://ftp.springer.de/pub/tex/latex/svmonot1/ (for monographs) and
   ftp://ftp.springer.de/pub/tex/latex/svmultt1/ (for summer schools/tutorials).

Additional technical instructions, if necessary, are available on request from lnm@springer.com.

4. Careful preparation of the manuscripts will help keep production time short besides ensuring satisfactory appearance of the finished book in print and online. After acceptance of the manuscript authors will be asked to prepare the final LaTeX source files and also the corresponding dvi-, pdf- or zipped ps-file. The LaTeX source files are essential for producing the full-text online version of the book (see http://www.springerlink.com/openurl.asp?genre=journal&issn=0075-8434 for the existing online volumes of LNM). The actual production of a Lecture Notes volume takes approximately 12 weeks.

5. Authors receive a total of 50 free copies of their volume, but no royalties. They are entitled to a discount of 33.3 % on the price of Springer books purchased for their personal use, if ordering directly from Springer.

6. Commitment to publish is made by letter of intent rather than by signing a formal contract. Springer-Verlag secures the copyright for each volume. Authors are free to reuse material contained in their LNM volumes in later publications: a brief written (or e-mail) request for formal permission is sufficient.

Addresses:
Professor J.-M. Morel, CMLA,
École Normale Supérieure de Cachan,
61 Avenue du Président Wilson, 94235 Cachan Cedex, France
E-mail: morel@cmla.ens-cachan.fr

Professor B. Teissier, Institut Mathématique de Jussieu,
UMR 7586 du CNRS, Équipe "Géométrie et Dynamique",
175 rue du Chevaleret
75013 Paris, France
E-mail: teissier@math.jussieu.fr

For the "Mathematical Biosciences Subseries" of LNM:

Professor P. K. Maini, Center for Mathematical Biology,
Mathematical Institute, 24-29 St Giles,
Oxford OX1 3LP, UK
E-mail: maini@maths.ox.ac.uk

Springer, Mathematics Editorial, Tiergartenstr. 17,
69121 Heidelberg, Germany,
Tel.: +49 (6221) 4876-8259

Fax: +49 (6221) 4876-8259
E-mail: lnm@springer.com